Statistics with Confidence

Statistics with Confidence

Confidence intervals and statistical guidelines

Edited by

MARTIN J GARDNER, PHD, FFOM
Professor of medical statistics, MRC Environmental Epidemiology Unit, University of Southampton

and

DOUGLAS G ALTMAN, BSC
Head, Medical Statistics Laboratory, Imperial Cancer Research Fund, London

Published by the British Medical Journal
Tavistock Square, London WC1H 9JR

First published 1989
Reprinted 1989
Reprinted 1990 (3 times)

British Library Cataloguing in Publication Data

Statistics with confidence.
1. Mathematics. Interval analysis
I. Gardner. M.J. (Martin, John), 1940– II. Altman, Douglas G. III. British medical. journal
519.4

ISBN 0-7279-0222-9

Printed in Great Britain by The Universities Press (Belfast) Ltd.

Contributors

Douglas G Altman, BSC, *head, Medical Statistics Laboratory, Imperial Cancer Research Fund, London WC2A 3PX*

Michael J Campbell, MSC, PHD, *senior lecturer in medical statistics, Medical Statistics and Computing (University of Southampton), Southampton General Hospital, Southampton SO9 4XY*

Martin J Gardner, PHD, FFOM, *professor of medical statistics, MRC Environmental Epidemiology Unit (University of Southampton), Southampton General Hospital, Southampton SO9 4XY*

Sheila M Gore, MA, PHD, *medical statistician, MRC Biostatistics Unit, Cambridge CB2 2BW*

David Machin,* MSC, PHD, *senior lecturer in medical statistics, Medical Statistics and Computing (University of Southampton), Southampton General Hospital, Southampton SO9 4XY*

Julie A Morris, MSC, *medical statistician, Department of Medical Statistics, Withington Hospital, West Didsbury, Manchester M20 8LR*

Stuart J Pocock, MSC, PHD, *professor of medical statistics, Department of Clinical Epidemiology and General Practice, Royal Free Hospital School of Medicine, London NW3 2PF*

* Now: *chief statistician, MRC Cancer Trials Office, Cambridge CB2 2BW.*

Computer program

A computer program, Confidence Interval Analysis (CIA), suitable for use on IBM or compatible microcomputers, has been specially prepared to carry out the calculations described in this book. For further information on availability and an order form write to: British Medical Journal (CIA), BMA House, Tavistock Square, London WC1H 9JR.

Contents

PART II STATISTICAL GUIDELINES AND CHECK LISTS

Source of contents

INTRODUCTION specially written for this book

Part I Estimation and confidence intervals

1 Gardner MJ, Altman DG. Estimating with confidence. *Br Med J* 1988;**296**:1210–1 (revised)
2 Gardner MJ, Altman DG. Confidence intervals rather than P values: estimation rather than hypothesis testing. *Br Med J* 1986;**292**:746–50 (revised)
3 From appendix 2 of source reference to chapter 2 (revised and expanded)
4 From appendix 2 of source reference to chapter 2 (revised and expanded)
5 Altman DG, Gardner MJ. Calculating confidence intervals for regression and correlation. *Br Med J* 1988;**296**:1238–42 (revised and expanded)
6 Morris JA, Gardner MJ. Calculating confidence intervals for relative risks (odds ratios) and standardised ratios and rates. *Br Med J* 1988;**296**:1313–6 (revised and expanded)
7 Machin D, Gardner MJ. Calculating confidence intervals for survival time analyses. *Br Med J* 1988;**296**:1369–71 (revised)
8 Campbell MJ, Gardner MJ. Calculating confidence intervals for some non-parametric analyses. *Br Med J* 1988;**296**:1454–6 (revised)

Part II Statistical guidelines and check lists

9 Altman DG, Gore SM, Gardner MJ, Pocock SJ. Statistical guidelines for contributors to medical journals. *Br Med J* 1983;**286**:1489–93 (revised)
10 Gardner MJ, Machin D, Campbell MJ. Use of check lists in assessing the statistical content of medical studies. *Br Med J* 1986;**292**:810–2 (revised)

Part III Tables specially prepared for this book

NOTATION specially written for this book

Acknowledgments

We thank the many people who kindly read and constructively criticised draft versions of the original papers; Drs JR Eason, MB McIllmurray, and G Dale for making available unpublished data; Dr Clive Osmond for calculating the values given in table 3 of part III from the formula for the Poisson distribution; Mr Stephen Gardner for helping to program the calculations to produce the values in tables 5 and 6 of part III; and Mrs Brigid Howells for her careful typing of several drafts of most of the chapters. We would also like to ackowledge the encouragement and collaboration of the editorial staff of the British Medical Journal.

Introduction

MARTIN J GARDNER, DOUGLAS G ALTMAN

This book collects together seven papers and a leader published recently in the "Statistics in Medicine" columns of the *British Medical Journal*. All have been revised and most have been expanded. The main methods and ideas considered are illustrated by medical examples but are, none the less, applicable to statistical practice in other disciplines.

Part I, which includes most of the chapters, covers the use of confidence intervals for the presentation of research findings. In this way results are given as estimates of effects based on the study sample together with an indication of their imprecision due to sampling variation, rather than in terms of the outcomes of null hypothesis tests and P values. The rationale for preferring the estimation and confidence interval approach is described in chapter 1 and, in more detail, in chapter 2.

Chapters 3 to 8 contain methods of calculating confidence intervals for most common medical applications, since these techniques are not always covered or easy to find in statistical textbooks. For some statistics, where exact calculation of the confidence intervals is difficult, a number of different approximate methods have been published. In these cases we have selected one that is relatively easy to use without a computer and which will not be misleading; situations are indicated where more complex approximate or exact methods are desirable. Worked examples, mainly using data taken from published papers, are used throughout. For those readers less familiar with statistical methods a section describing the mathematical notation used in these chapters is provided at the end of the book. Tables specially prepared for calculating most of the confidence intervals described in the text are given in part III. A computer program, CIA, is available to calculate the confidence intervals (for details see cover or p viii).

Chapter 9, in part II, contains statistical guidelines for authors preparing research papers for publication. These guidelines cover methods of presenting a wide range of the statistical aspects of studies. The recommendations are presented in order of the usual sections of Introduction, Methods, Results, and Discussion. Since their publication the guidelines have been endorsed by *Statistics in Medicine*[1] and are referred to in the *British Medical Journal* and other journals in their advice to authors.

Chapter 10, also in part II, considers check lists that are used regularly by statistical assessors of manuscripts submitted to the *British Medical Journal*. The check lists are also designed to help medical researchers and writers plan and analyse their studies and to prepare papers for publication. The check lists have recently been adopted also by the *Medical Journal of Australia*.[2]

Clearly this book is not intended as a comprehensive statistical textbook and for further details of statistical methods the reader is referred to other sources.[3–6] Additionally, in each chapter specific appropriate references are given.

1 Johnson T. Editorial: statistical guidelines for medical journals. *Statistics in Medicine* 1984; **3**:97–9.
2 Berry G. Statistical guidelines and statistical guidance. *Med J Aust* 1987;**146**:408–9.
3 Armitage P, Berry, G. *Statistical methods in medical research*. 2nd ed. Oxford: Blackwell, 1987.
4 Bland M. *An introduction to medical statistics*. Oxford: University Press, 1987.
5 Bradford Hill A. *A short textbook of medical statistics*. 11th ed. London: Hodder and Stoughton, 1984.
6 Colton T. *Statistics in medicine*. Boston: Little, Brown, 1974.

Part I
Estimation and confidence intervals

1
Estimating with confidence

MARTIN J GARDNER, DOUGLAS G ALTMAN

Statistical analysis of medical studies is based on the key idea that we make observations on a sample of subjects and then draw inferences about the population of all such subjects from which the sample is drawn. If the study sample is not representative of the population we may well be misled and statistical procedures cannot help. But even a well designed study can give only an idea of the answer sought because of random variation in the sample. Thus results from a single sample are subject to statistical uncertainty, which is strongly related to the size of the sample. Examples of the statistical analysis of sample data would be calculating the difference between the proportions of patients improving on two treatment regimens or the slope of the regression line relating two variables. These quantities will be imprecise estimates of the values in the overall population, but fortunately the imprecision can itself be estimated and incorporated into the presentation of findings. Presenting study findings directly on the scale of original measurement, together with information on the inherent imprecision due to sampling variability, has distinct advantages over just giving P values usually dichotomised into "significant" or "non-significant." This is the rationale for using confidence intervals.

The main purpose of confidence intervals is to indicate the (im)precision of the sample study estimates as population values. Consider the following points for example: a difference of 20% between the percentages improving in two groups of 80 patients having treatments A and B was reported, with a 95% confidence interval of 6% to 34% (see chapter 4). Firstly, a possible difference in treatment effectiveness of less than 6% or of more than 34% is not excluded by such values being outside the confidence interval—they are simply less likely than those inside the confidence interval. Secondly, the middle half of the confidence interval (13% to 27%)

is more likely to contain the population value than the extreme two quarters (6% to 13% and 27% to 34%)—in fact the middle half forms a 67% confidence interval. Thirdly, regardless of the width of the confidence interval, the sample estimate is the best indicator of the population value—in this case a 20% difference in treatment response.

The *British Medical Journal* now expects scientific papers submitted to it to contain confidence intervals when appropriate.[1] It also wants a reduced emphasis on the presentation of P values from hypothesis testing (see chapter 2). *The Lancet*,[2 3] the *Medical Journal of Australia*,[4] the *American Journal of Public Health*,[5] and the *British Heart Journal*[6] have implemented the same policy, and it has been endorsed by the International Committee of Medical Journal Editors.[7] One of the blocks to implementing the policy has been that the methods needed to calculate confidence intervals are not readily available in most statistical textbooks. The chapters that follow present appropriate techniques for most common situations. Further articles in the *American Journal of Public Health* and the *Annals of Internal Medicine* have debated the uses of confidence intervals and hypothesis tests and discussed the interpretation of confidence intervals.[8–14]

So when should confidence intervals be calculated and presented? Essentially confidence intervals become relevant whenever an inference is to be made from the study results to the wider world. Such an inference will relate to summary, not individual, characteristics—for example, rates, differences in medians, regression coefficients, etc. The calculated interval will give us a range of values within which we can have a chosen confidence of it containing the population value. The most usual degree of confidence presented is 95%, but any suggestion to standardise on 95%[2 3] would not seem desirable.[15]

Thus, a single study usually gives an imprecise sample estimate of the overall population value in which we are interested. This imprecision is indicated by the width of the confidence interval: the wider the interval the less precision. The width depends essentially on three factors. Firstly, the sample size: larger sample sizes will give more precise results with narrower confidence intervals (see chapter 2). In particular, wide confidence intervals emphasise the unreliability of conclusions based on small samples. Secondly, the variability of the characteristic being studied: the less variable it is (between subjects, within subjects, from measurement error, and from other sources) the more precise the sample estimate and the narrower the confidence interval. Thirdly, the degree of confidence required: the more confidence the wider the interval.

1 Langman MJS. Towards estimation and confidence intervals. *Br Med J* 1986;**292**:716.
2 Anonymous. Report with confidence [Editorial]. *Lancet* 1987;**i**:488.
3 Bulpitt CJ. Confidence intervals. *Lancet* 1987;**i**:494–7.
4 Berry G. Statistical significance and confidence intervals. *Med J Aust* 1986;**144**:618–9.
5 Rothman KJ, Yankauer A. Confidence intervals vs significance tests: quantitative interpretation (Editors' note). *Am J Public Health* 1986;**76**:587–8.
6 Evans SJW, Mills P, Dawson J. The end of the P value? *Br Heart J* 1988; **60**:177–80.
7 International Committee of Medical Journal Editors. Uniform requirements for manuscripts submitted to biomedical journals. *Br Med J* 1988;**296**:401–5.
8 DeRouen TA, Lachenbruch PA, Clark VA, *et al.* Four comments received on statistical testing and confidence intervals. *Am J Public Health* 1987;**77**:237–8.
9 Editor. Four comments received on statistical testing and confidence intervals. *Am J Public Health* 1987;**77**:238.
10 Thompson WD. Statistical criteria in the interpretation of epidemiological data. *Am J Public Health* 1987;**77**:191–4.
11 Thompson WD. On the comparison of effects. *Am J Public Health* 1987;**77**:491–2.
12 Poole C. Beyond the confidence interval. *Am J Public Health* 1987;**77**:195–9.
13 Poole C. Confidence intervals exclude nothing. *Am J Public Health* 1987;**77**:492–3.
14 Braitman, LE. Confidence intervals extract clinically useful information from data. *Ann Intern Med* 1988;**108**:296–8.
15 Gardner MJ, Altman DG. Using confidence intervals. *Lancet* 1987;**i**:746.

2

Estimation rather than hypothesis testing: confidence intervals rather than P values

MARTIN J GARDNER, DOUGLAS G ALTMAN

Summary

Overemphasis on hypothesis testing—and the use of P values to dichotomise significant or non-significant results—has detracted from more useful approaches to interpreting study rcsults, such as estimation and confidence intervals. In medical studies investigators should usually be interested in determining the size of difference of a measured outcome between groups, rather than a simple indication of whether or not it is statistically significant. Confidence intervals present a range of values, on the basis of the sample data, in which the population value for such a difference may lie.

Confidence intervals, if appropriate to the type of study, should be used for major findings in both the main text of a paper and its abstract.

Introduction

Over the past two or three decades the use of statistics in medical journals has increased tremendously. One unfortunate consequence has been a shift in emphasis away from the basic results towards an undue concentration on hypothesis testing. In this approach data are examined in relation to a statistical "null" hypothesis, and the practice has led to the mistaken belief that studies should aim at obtaining "statistical significance." On the contrary, the purpose of most research investigations in medicine is to determine the magnitude of some factor(s) of interest.

For example, a laboratory based study may investigate the difference in mean concentrations of a blood constituent between patients with and without a certain illness, while a clinical study may assess the difference in prognosis of patients with a particular disease treated by alternative regimens in terms of rates of cure, remission, relapse, survival, etc. The difference obtained in such a study will be only an estimate of what we really need, which is the result that would have been obtained had all the eligible subjects (the "population") been investigated rather than just a sample of them. What authors and readers should want to know is by how much the illness modified the mean blood concentrations or by how much the new treatment altered the prognosis, rather than only the level of statistical significance.

The excessive use of hypothesis testing at the expense of other ways of assessing results has reached such a degree that levels of significance are often quoted alone in the main text and abstracts of papers, with no mention of actual concentrations, proportions, etc, or their differences. The implication of hypothesis testing—that there can always be a simple "yes" or "no" answer as the fundamental result from a medical study—is clearly false and used in this way hypothesis testing is of limited value (see chapter 9).

We discuss here the rationale behind an alternative statistical approach—the use of confidence intervals; these are more informative than P values, and we recommend them for papers published in the *British Medical Journal*, which now requires them where appropriate, and more generally. This should not be taken to mean that confidence intervals should appear in all papers; in some cases, such as where the data are purely descriptive, confidence intervals are inappropriate and in others techniques for obtaining them are complex or unavailable.

Presentation of study results: limitations of P values

The common simple statements "$P<0{\cdot}05$," "$P>0{\cdot}05$," or "P = NS" convey little information about a study's findings and rely on an arbitrary convention of using the 5% level of statistical significance to define two alternative outcomes—significant or not significant—which is not helpful and encourages lazy thinking. Furthermore, even precise P values convey nothing about the sizes of the differences between study groups. Rothman pointed this out in 1978 and advocated the use of confidence intervals,[1] and recently he and his colleagues repeated the proposal.[2]

Presenting P values alone can lead to their being given more merit than they deserve. In particular, there is a tendency to equate statistical significance with medical importance or biological relevance. But small differences of no real interest can be statistically significant with large sample sizes, whereas clinically important effects may be statistically non-significant only because the number of subjects studied was small.

Presentation of study results: confidence intervals

It is more useful to present sample statistics as estimates of results that would be obtained if the total population were studied. The lack of precision of a sample statistic—for example, the mean—which results from both the degree of variability in the factor being investigated and the limited size of the study, can be shown advantageously by a confidence interval.

A confidence interval produces a move from a single value estimate—such as the sample mean, difference between sample means, etc—to a range of values that are considered to be plausible for the population. The width of a confidence interval associated with a sample statistic depends partly on its standard error, and hence on both the standard deviation and the sample size (see appendix 1 for a brief description of the important, but often misunderstood, distinction between the standard deviation and standard error). It also depends on the degree of "confidence" that we want to associate with the resulting interval.

Suppose that in a study comparing samples of 100 diabetic and 100 non-diabetic men of a certain age a difference of 6·0 mm Hg was found between their mean systolic blood pressures and that the standard error of this difference between sample means was 2·5 mm Hg—comparable to the difference between means in the Framingham study.[3] The 95% confidence interval for the population difference between means is from 1·1 to 10·9 mm Hg and is shown in fig 2.1 together with the original data. Details of how to calculate the confidence interval are given in chapter 3.

Put simply, this means that there is a 95% chance that the indicated range includes the "population" difference in mean blood pressure levels—that is, the value which would be obtained by including the total populations of diabetics and non-diabetics at

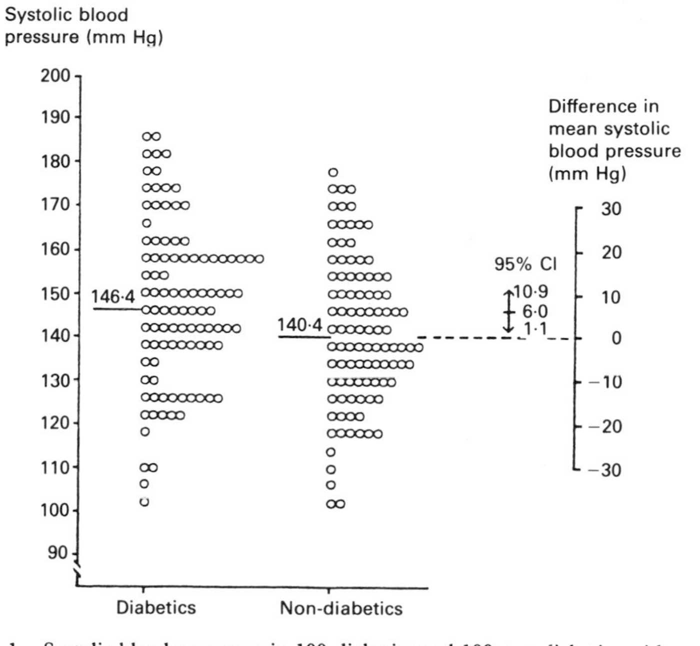

FIG 2.1—Systolic blood pressures in 100 diabetics and 100 non-diabetics with mean levels of 146·4 and 140·4 mm Hg respectively. The difference between the sample means of 6·0 mm Hg is shown to the right together with the 95% confidence interval from 1·1 to 10·9 mm Hg.

which the study is aimed. More exactly, in a statistical sense, the confidence interval means that if a series of identical studies were carried out repeatedly on different samples from the same populations, and a 95% confidence interval for the difference between the sample means calculated in each study, then, in the long run, 95% of these confidence intervals would include the population difference between means.

The sample size affects the size of the standard error and this in turn affects the width of the confidence interval. This is shown in fig 2.2, which shows the 95% confidence interval from samples with the same means and standard deviations as before but only half as large—that is, 50 diabetics and 50 non-diabetics. Reducing the sample size leads to less precision and an increase in the width of the confidence interval, in this case by some 40%.

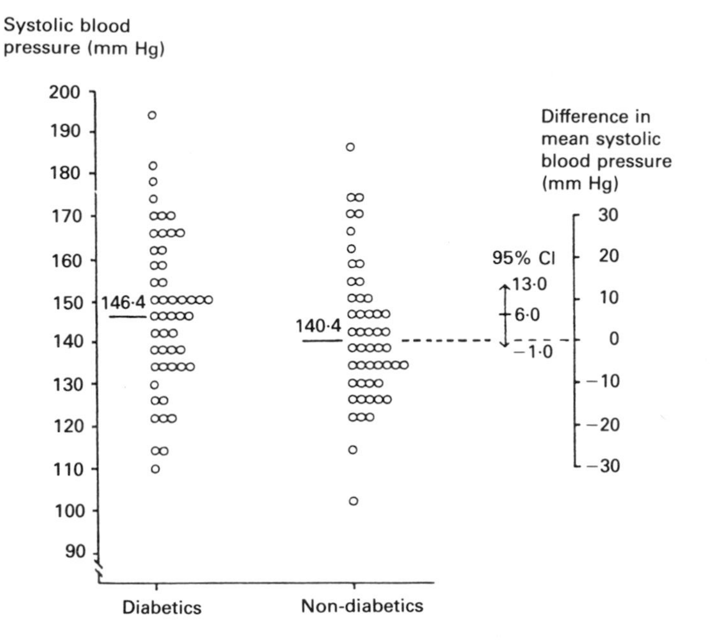

FIG 2.2—As fig 2.1 but showing results from two samples of half the size—that is, 50 subjects each. The means and standard deviations are as in fig 2.1, but the 95% confidence interval is wider, from −1·0 to 13·0 mm Hg, owing to the smaller sample sizes.

The investigator can select the degree of confidence associated with a confidence interval, though 95% is the most common choice—just as a 5% level of statistical significance is widely used. If greater or less confidence is required different intervals can be constructed: 99%, 95%, and 90% confidence intervals for the data in fig 2.1 are shown in fig 2.3. As would be expected, greater confidence that the population difference is within a confidence interval is obtained with wider intervals. In practice, intervals other than 99%, 95%, or 90% are rarely quoted. Appendix 2 shows the general method for calculating a confidence interval appropriate for most of the methods described in this book.

Confidence intervals convey only the effects of sampling variation on the precision of the estimated statistics and cannot control for non-sampling errors such as biases in design, conduct, or analysis.

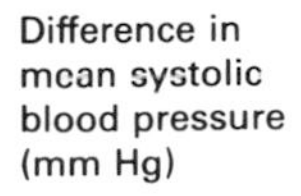

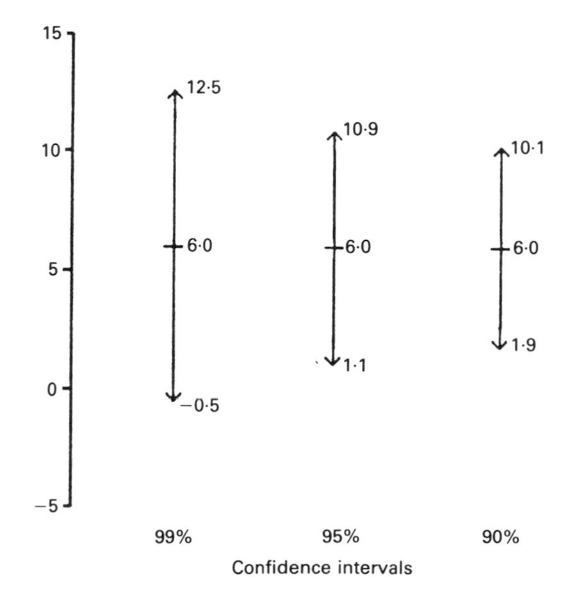

FIG 2.3—Confidence intervals associated with differing degrees of "confidence" using the same data as in fig 2.1.

Sample sizes and confidence intervals

In general, increasing the sample size will reduce the width of the confidence interval. If we assume the same means and standard deviations as in the example fig 2.4 shows the resulting 99%, 95%, and 90% confidence intervals for the difference in mean blood pressures for sample sizes of up to 500 in each group. The benefit, in terms of narrowing the confidence interval, of a further increase in the number of subjects falls sharply with increasing sample size. Similar effects occur in estimating other statistics such as proportions.

For a total study sample size of N subjects, the confidence interval for the difference in population means is narrowest when both groups are of size $N/2$. However, the width of the confidence

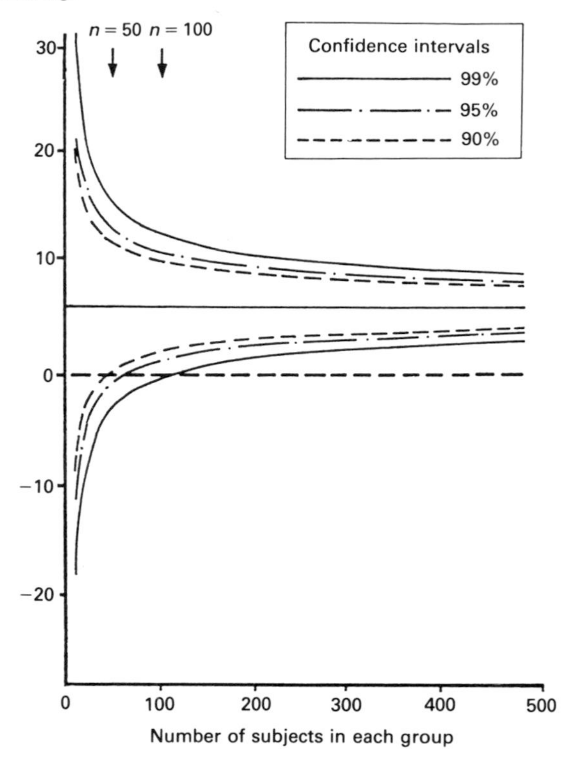

FIG 2.4—Confidence intervals resulting from the same means and standard deviations as in fig 2.1 but showing the effect on the confidence interval of sample sizes of up to 500 subjects in each group. The two horizontal lines show: – – – zero difference between means, ——— study difference between means of 6·0 mm Hg. The arrows indicate the confidence intervals shown in figs 2.1–2.3 for sample sizes of 100 and 50 in each group.

interval will be only slightly larger with differing numbers in each group unless one group size is relatively small.

During the planning stage of a study it is possible to estimate the sample size that should be used by stating the width of the confidence interval required at the end of the study and carrying out the appropriate calculation (see, for example, ref 4).

Confidence intervals and statistical significance

There is a close link between the use of a confidence interval and a two sided hypothesis test. If the confidence interval is calculated then the result of the hypothesis test (often less accurately referred to as a significance test) can be inferred at an associated level of statistical significance. The right hand scale in fig 2.1 includes the point that represents a zero difference in mean blood pressure between diabetics and non-diabetics. This zero difference between means corresponds to the value examined under the "null hypothesis" and, as fig 2.1 shows, it is outside the 95% confidence interval. This indicates that a statistically significant difference between the sample means at the 5% level would result from applying the appropriate unpaired *t* test. Figure 2.3, however, shows that the P value is greater than 1% because zero is inside the 99% confidence interval, so $0{\cdot}01 < P < 0{\cdot}05$. By contrast, had zero been within the 95% confidence interval this would have indicated a non-significant result at the 5% level. Such an example is shown in fig 2.2 for the smaller samples.

The 95% confidence interval covers a wide range of possible population mean differences, even though the sample difference between means is different from zero at the 5% level of statistical significance. In particular, the 95% confidence interval shows that the study result is compatible with a small difference of around 1 mm Hg as well as with a difference as great as 10 mm Hg in mean blood pressures. Nevertheless, the difference between population means is much more likely to be near to the middle of the confidence interval than towards the extremes. Although the confidence interval is wide, the best estimate of the population difference is 6·0 mm Hg, the difference between the sample means.

This example therefore shows the lack of precision of the observed sample difference between means as an estimate of the population value, and this is clear in each of the three confidence intervals shown in fig 2.3. It also shows the weakness of considering statistical significance in isolation from the numerical estimates.

The confidence interval thus provides a range of possibilities for the population value, rather than an arbitrary dichotomy based solely on statistical significance. It conveys more useful information at the expense of precision of the P value. However, the actual P value is helpful in addition to the confidence interval, and preferably both should be presented. If one has to be excluded however, it should be the P value.

Suggested mode of presentation

In content, our only proposed change is that confidence intervals should be reported instead of standard errors. This will encourage a move away from the current emphasis on statistical significance. For the major finding(s) of a study we recommend that full statistical information should be given, including sample estimates, confidence intervals, test statistics, and P values—assuming that basic details, such as sample sizes and standard deviations, have been reported earlier in the paper. The major findings would include at least those related to the original hypothesis(es) of the study and those reported in the abstract.

For the above example the textual presentation of the results might read:

> The difference between the sample mean systolic blood pressures in diabetics and non-diabetics was 6·0 mm Hg, with a 95% confidence interval from 1·1 to 10·9 mm Hg; the *t* test statistic was 2·4, with 198 degrees of freedom and an associated P value of 0·02.

In short:

> Mean 6·0 mm Hg, 95% confidence interval 1·1 to 10·9; $t = 2{\cdot}4$, df = 198, P = 0·02.

It is preferable to use the word "to" for separating the two values rather than a dash, as a dash is confusing when at least one of the numbers is negative. The use of the ± sign should also be avoided (see appendix 1 and chapter 9). The exact P value from the *t* distribution is 0·017 32, but one or two significant figures are enough (see chapter 9); this value is seen to be within the range 0·01 to 0·05 determined earlier from the confidence intervals. Often a range for P will need to be given because only limited figures are available in published tables—for example, $0{\cdot}3 < P < 0{\cdot}4$.

The two extremes of a confidence interval are sometimes presented as confidence limits. However, the word "limits" suggests that there is no going beyond and may be misunderstood because, of course, the population value will not always lie within the confidence interval. Moreover, there is a danger that one or other of the "limits" will be quoted in isolation from the rest of the results, with misleading consequences. For example, concentrating only on the larger limit and ignoring the rest of the confidence interval would misrepresent the finding by exaggerating the study difference. Conversely, quoting only the smaller limit would incorrectly underestimate the difference. The confidence interval is thus preferable because it focuses on the range of values.

The same notation can be used for presenting confidence intervals in tables. Thus, a column headed "95% confidence interval" or "95% CI" would have rows of intervals: 1·1 to 10·9, 48 to 85, etc. Confidence intervals can also be incorporated into figures, where they are preferable to the widely used standard error, which is often shown solely in one direction from the sample estimate. If individual data values can be shown as well, which is usually possible for small samples, this is even more informative. Thus in fig 2.1, despite the considerable overlap of the two sets of sample data, the shift in means is shown by the 95% confidence interval excluding zero. For paired samples the individual differences can be plotted advantageously in a diagram.

The example given here of the difference between two means is common. Although there is some intrinsic interest in the mean values themselves, inferences from a study will be concerned mainly with their difference. Giving confidence intervals for each mean separately is therefore unhelpful, because these do not indicate the precision of the difference or its statistical significance.[5 6] Thus, the major contrasts of a study should be shown directly, rather than only vaguely in terms of the separate means (or proportions).

For a paper with only a limited number of statistical comparisons related to the initial hypotheses confidence intervals are recommended throughout. Where multiple comparisons are concerned, however, the usual problems of interpretation arise, since some confidence intervals will exclude the "null" value—for example, zero difference—through random sampling variation alone. This mirrors the situation of calculating a multiplicity of P values, where not all statistically significant differences are likely to represent real effects.[7] Judgment needs to be exercised over the number of statistical comparisons made, with confidence intervals and P values calculated, to avoid misleading both authors and readers (see chapter 9).

Conclusion

The excessive use of hypothesis testing at the expense of more informative approaches to data interpretation is an unsatisfactory way of assessing and presenting statistical findings from medical studies. We prefer the use of confidence intervals, which present the results directly on the scale of data measurement. We have also suggested a notation for confidence intervals which is intended to force clarity of meaning.

Confidence intervals, which also have a link to the outcome of hypothesis tests, should become the standard method for presenting the statistical results of major findings.

Appendix 1: Standard deviation and standard error

When numerical findings are reported, regardless of whether or not their statistical significance is quoted, they are often presented with additional statistical information. The distinction between two widely quoted statistics—the standard deviation and the standard error—is, however, often misunderstood.[8–13]

The standard deviation is a measure of the variability between individuals in the level of the factor being investigated, such as blood alcohol concentrations in a sample of car drivers, and is thus a descriptive index. By contrast, the standard error is a measure of the uncertainty in a sample statistic. For example, the standard error of the mean indicates the uncertainty of the mean blood alcohol concentration among the sample of drivers as an estimate of the mean value among the population of all car drivers. The standard deviation is relevant when variability between individuals is of interest; the standard error is relevant to summary statistics such as means, proportions, differences, regression slopes, etc (see chapter 9).

The standard error of the sample statistic, which depends on both the standard deviation and the sample size, is a recognition that a sample is most unlikely to determine the population value exactly. In fact, if a further sample is taken in identical circumstances almost certainly it will produce a different estimate of the same population value. The sample statistic is therefore imprecise, and the standard error is a measure of this imprecision. By itself the standard error has limited meaning, but it can be used to produce a confidence interval, which does have a useful interpretation.

In many publications a ± sign is used to join the standard deviation (SD) or standard error (SE) to an observed mean—for example, $69{\cdot}4 \pm 9{\cdot}3$ kg—but the notation gives no indication whether the second figure is the standard deviation or the standard error (or something else).[13] As is suggested in chapter 9, a clearer presentation would be in the unambiguous form "the mean was 69·4 kg (SD 9·3 kg)." The present policy of the *British Medical Journal* and many other journals is to remove ± signs and request authors to indicate clearly whether the standard deviation or standard error is being quoted. All journals should follow this practice;[14] it avoids any possible misunderstanding from the omission of SD or SE.[15]

Appendix 2: Constructing confidence intervals

Frequently we can reasonably assume that the estimate of interest, such as the difference between two proportions (see chapter 4), has a Normal sampling distribution. To construct a confidence interval for the population value we are interested in the range of values within which the sample estimate would fall on most occasions (see main text of this chapter). The calculation of confidence intervals is simplified by the ability to convert the standard Normal distribution into the Normal distribution of interest by multiplying by the standard error of the estimate and adding the value of the estimate.

To construct a 95% confidence interval, say, we use the central 95% of the standard Normal distribution, so we need the values that cut off $2\frac{1}{2}$% of the distribution at each end (or "tail"). Thus we need the values of $N_{0\cdot025}$ and $N_{0\cdot975}$. In general we construct a $100(1-\alpha)$% confidence interval using the values $N_{\alpha/2}$ and $N_{1-\alpha/2}$, which cut off the bottom and top $100\alpha/2$% of the distribution. Figure 2.5 illustrates the procedure.

To convert back to the scale of the original data we multiply the two values $N_{\alpha/2}$ and $N_{1-\alpha/2}$ by the standard error (SE) and add them to the estimate, to get the $100(1-\alpha)$% confidence interval as

$$\text{estimate} + (N_{\alpha/2} \times \text{SE}) \quad \text{to} \quad \text{estimate} + (N_{1-\alpha/2} \times \text{SE}).$$

As explained in the notation list at the end of the book, an equivalent expression is

$$\text{estimate} - (N_{1-\alpha/2} \times \text{SE}) \quad \text{to} \quad \text{estimate} + (N_{1-\alpha/2} \times \text{SE}),$$

which makes more explicit the symmetry of the confidence interval around the estimate.

For some estimates, such as the difference between sample means (see chapter 3), the appropriate sampling distribution is the *t*

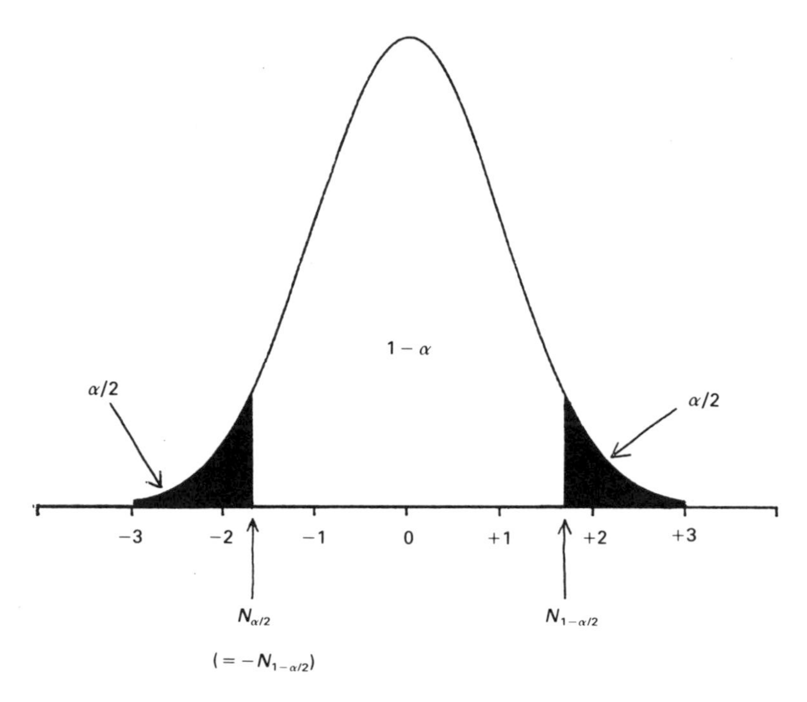

FIG 2.5—Standard Normal distribution curve.

distribution. Exactly the same procedure is adopted but with $N_{1-\alpha/2}$ replaced by $t_{1-\alpha/2}$. For other estimates different sampling distributions are relevant, such as the Poisson distribution for standardised mortality ratios (see chapter 6) or the Binomial for medians (see chapter 8).

1 Rothman K. A show of confidence. *N Engl J Med* 1978;**299**:1362–3.
2 Poole C, Lanes S, Rothman KJ. Analysing data from ordered categories. *N Engl J Med* 1984;**311**:1382.
3 Kannel WB, McGee DL. Diabetes and cardiovascular risk factors: the Framingham study. *Circulation* 1979;**59**:8–13.
4 Armitage P, Berry G. *Statistical methods in medical research.* 2nd ed. Oxford: Blackwell, 1987:180.
5 Browne RH. On visual assessment of the significance of a mean difference. *Biometrics* 1979;**35**:657–65.
6 Altman DG. Statistics and ethics in medical research: VI—presentation of results. *Br Med J* 1980;**281**:1542–4.
7 Jones, DR, Rushton L. Simultaneous inference in epidemiological studies. *Int J Epidemiol* 1982;**11**:276–82.
8 Gardner MJ. Understanding and presenting variation. *Lancet* 1975;**i**:230–1.
9 Feinstein AR. Clinical biostatistics XXXVII: demeaned errors, confidence games, non-

plussed minuses, inefficient coefficients, and other statistical disruptions of scientific communication. *Clin Pharmacol Ther* 1976;**20**:617–31.

10 Bunce H, Hokanson JA, Weiss GB. Avoiding ambiguity when reporting variability in biomedical data. *Am J Med* 1980;**69**:8–9.

11 Altman DG. Statistics in medical journals. *Statistics in Medicine* 1982;**1**:59–71.

12 Brown GW. Standard deviation, standard error: which "standard" should we use? *Am J Dis Child* 1982;**136**:937–41.

13 Altman DG, Gardner MJ. Presentation of variability. *Lancet* 1986;**ii**:639.

14 Huth EJ. Uniform requirements for manuscripts: the new, third edition. *Ann Intern Med* 1988;**108**:298–9.

15 Bailar JC, Mosteller F. Guidelines for statistical reporting in articles for medical journals: amplifications and explanations. *Ann Intern Med* 1988; **108**:266–73.

3

Calculating confidence intervals for means and their differences

DOUGLAS G ALTMAN, MARTIN J GARDNER

The rationale behind the use of confidence intervals was described in chapters 1 and 2. Here formulae for calculating confidence intervals are given for means and their differences. There is a common underlying principle of subtracting and adding to the sample statistic a multiple of its standard error (SE). This extends to other statistics, such as proportions and regression coefficients, but is not universal.

Confidence intervals for means are constructed using the t distribution if the data have an approximately Normal distribution. For differences between two means the data should also have similar standard deviations (SDs) in each study group. This is implicit in the example given in chapter 2 and in the worked examples below. The calculations have been carried out to full arithmetical precision, as is recommended practice (see chapter 9), but intermediate steps are shown as rounded results.

The case of non-Normal data is discussed both in this chapter and in chapter 8.

A confidence interval indicates the precision of the sample mean or the difference between two sample means as an estimate of the overall population value. As such, confidence intervals convey the effects of sampling variation but cannot control for non-sampling errors in study design or conduct.

Single sample

The confidence interval for a population mean is derived using the mean ($\bar{x}$) and its standard error from a sample of size n. For this case the $SE = SD/\sqrt{n}$. Thus, the confidence interval is given by

$$\bar{x} - (t_{1-\alpha/2} \times SE) \quad \text{to} \quad \bar{x} + (t_{1-\alpha/2} \times SE),$$

where $t_{1-\alpha/2}$ is the appropriate value from the t distribution with $n-1$ degrees of freedom associated with a "confidence" of $100(1-\alpha)\%$. For a 95% confidence interval α is 0·05, for a 99% confidence interval α is 0·01, and so on. Values of t can be found from table 2 in part III, in statistical textbooks, or from *Geigy Scientific Tables*.[1] For a 95% confidence interval the value of t will be close to 2·0 for samples of 20 upwards but noticeably greater than 2·0 for smaller samples.

Worked example

Blood pressure levels were measured in a sample of 100 diabetic men aged 40–49 years. The mean systolic blood pressure was 146·4 mm Hg and the standard deviation 18·5 mm Hg. The standard error of the mean is thus found as $18{\cdot}5/\sqrt{100} = 1{\cdot}85$ mm Hg.

To calculate the 95% confidence interval the appropriate value of $t_{0{\cdot}975}$ with 99 degrees of freedom is 1·984. The 95% confidence interval for the population value of the mean systolic blood pressure is then given by

$$146{\cdot}4 - (1{\cdot}984 \times 1{\cdot}85) \quad \text{to} \quad 146{\cdot}4 + (1{\cdot}984 \times 1{\cdot}85)$$

that is, from 142·7 to 150·1 mm Hg.

Two samples: unpaired case

The confidence interval for the difference between two population means is derived in a similar way. Suppose $\bar{x}_1$ and $\bar{x}_2$ are the two sample means, s_1 and s_2 the corresponding standard deviations, and n_1 and n_2 the sample sizes. Firstly, we need a "pooled" estimate of the standard deviation, which is given by

$$s = \sqrt{\frac{(n_1-1)s_1^2 + (n_2-1)s_2^2}{n_1+n_2-2}}.$$

From this the standard error of the difference between the two sample means is

$$\mathrm{SE}_{\mathrm{diff}} = s \times \sqrt{\frac{1}{n_1} + \frac{1}{n_2}}.$$

The $100(1-\alpha)\%$ confidence interval for the difference in the two population means is then

$$\bar{x}_1 - \bar{x}_2 - (t_{1-\alpha/2} \times \mathrm{SE}_{\mathrm{diff}}) \quad \text{to} \quad \bar{x}_1 - \bar{x}_2 + (t_{1-\alpha/2} \times \mathrm{SE}_{\mathrm{diff}}),$$

where $t_{1-\alpha/2}$ is taken from the t distribution with n_1+n_2-2 degrees of freedom (see table 2 in part III).

If the standard deviations differ considerably then a common pooled estimate is not appropriate unless a suitable transformation of scale can be found. Otherwise obtaining a confidence interval is more complex.[2]

Worked example

Blood pressure levels were measured in 100 diabetic and 100 non-diabetic men aged 40–49 years. Mean systolic blood pressures were 146·4 mm Hg (SD 18·5) among the diabetics and 140·4 mm Hg (SD 16·8) among the non-diabetics, giving a difference between sample means of 6·0 mm Hg.

Using the formulas given above the pooled estimate of the standard deviation is

$$s = \sqrt{\frac{(99 \times 18{\cdot}5^2) + (99 \times 16{\cdot}8^2)}{198}} = 17{\cdot}7 \text{ mm Hg},$$

and the standard error of the difference between the sample means is

$$\mathrm{SE}_{\mathrm{diff}} = 17{\cdot}7 \times \sqrt{\frac{1}{100} + \frac{1}{100}} = 2{\cdot}50 \text{ mm Hg}.$$

To calculate the 95% confidence interval the appropriate value of $t_{0{\cdot}975}$ with 198 degrees of freedom is 1·972. Thus the 95% confidence interval for the difference in population means is given by

$$6{\cdot}0 - (1{\cdot}972 \times 2{\cdot}50) \quad \text{to} \quad 6{\cdot}0 + (1{\cdot}972 \times 2{\cdot}50)$$

that is, from 1·1 to 10·9 mm Hg, as shown in fig 2.1.

Suppose now that the samples had been of only 50 men each but that the means and standard deviations had been the same. Then the pooled standard deviation would remain 17·7 mm Hg, but the standard error of the difference between the sample means would become

$$\mathrm{SE}_{\mathrm{diff}} = 17{\cdot}7 \times \sqrt{\frac{1}{50} + \frac{1}{50}} = 3{\cdot}53 \text{ mm Hg}.$$

The appropriate value of $t_{0{\cdot}975}$ on 98 degrees of freedom is 1·984, and the 95% confidence interval is calculated as

$$6{\cdot}0 - (1{\cdot}984 \times 3{\cdot}53) \quad \text{to} \quad 6{\cdot}0 + (1{\cdot}984 \times 3{\cdot}53)$$

that is, from −1·0 to 13·0 mm Hg, as shown in fig 2.2.

For the original samples of 100 each the appropriate values of $t_{0{\cdot}995}$ and $t_{0{\cdot}95}$ with 198 degrees of freedom to calculate the 99% and 90% confidence intervals are 2·601 and 1·653, respectively. Thus the 99% confidence interval is calculated as

$$6{\cdot}0-(2{\cdot}601\times 2{\cdot}50) \quad \text{to} \quad 6{\cdot}0+(2{\cdot}601\times 2{\cdot}50)$$

that is, from $-0{\cdot}5$ to 12·5 mm Hg (fig 2.3), and the 90% confidence interval is given by

$$6{\cdot}0-(1{\cdot}653\times 2{\cdot}50) \quad \text{to} \quad 6{\cdot}0+(1{\cdot}653\times 2{\cdot}50)$$

that is, from 1·9 to 10·1 mm Hg (fig 2.3).

Two samples: paired case

Paired data arise in studies of repeated measurements—for example, at different times or in different circumstances on the same subjects—and matched case-control comparisons. For such data thc same formulae as for the single sample case are used to calculate the confidence interval, where $\bar{x}$ and SD are now the mean and standard deviation of the individual within subject or patient-control differences.

Worked example

Systolic blood pressure levels were measured in 16 middle aged men before and after a standard exercise, giving the results shown in table 3.1.

TABLE 3.1—Systolic blood pressure levels (mm Hg) in 16 men before and after exercise

Subject number	Systolic blood pressure (mm Hg)		Difference
	Before	After	After–before
1	148	152	+4
2	142	152	+10
3	136	134	−2
4	134	148	+14
5	138	144	+6
6	140	136	−4
7	132	144	+12
8	144	150	+6
9	128	146	+18
10	170	174	+4
11	162	162	0
12	150	162	+12
13	138	146	+8
14	154	156	+2
15	126	132	+6
16	116	126	+10

The mean difference (rise) in systolic blood pressure following exercise was 6·6 mm Hg. The standard deviation of the differences, shown in the last column of table 3.1, is 6·0 mm Hg. Thus the standard error of the mean difference is found as $6{\cdot}0/\sqrt{16} = 1{\cdot}49$ mm Hg.

To calculate the 95% confidence interval the appropriate value of $t_{0{\cdot}975}$ with 15 degrees of freedom is 2·131. The 95% confidence interval for the population value of the mean systolic blood pressure increase after the standard exercise is then given by

$$6{\cdot}6 - (2{\cdot}131 \times 1{\cdot}49) \quad \text{to} \quad 6{\cdot}6 + (2{\cdot}131 \times 1{\cdot}49)$$

that is, from 3·4 to 9·8 mm Hg.

Non-Normal data

The sample data may have to be transformed on to a different scale to achieve approximate Normality. The most common reason is because the distribution of the observations is skewed, with a long "tail" of high values. The logarithmic transformation is the most frequently used. Transformation often also helps to make the standard deviations on the transformed scale in different groups more similar.

Single sample

For a single sample a mean and confidence interval can be constructed from the transformed data and then transformed back to the original scale of measurement. This is preferable to presenting the results in units of, say, log mm Hg. With highly skewed or otherwise awkward data the median may be preferable to the mean as a measure of central tendency and used with non-parametric methods of analysis. Confidence intervals can be calculated for the median (see chapter 8).

Worked example

Table 3.2 shows T4 and T8 lymphocyte counts in 28 haemophiliacs[3] ranked in increasing order of the T4 counts.

Suppose that we wish to calculate a confidence interval for the mean T4 lymphocyte count in the population of haemophiliacs. Inspection of histograms and plots of the data reveals that whereas the distribution of T4 values is skewed, after logarithmic transformation the values of $\log_e(\text{T4})$ have a symmetric near Normal distribution. We can thus apply the method given previously for calculating a confidence interval for a population mean derived from a single sample of observations.

TABLE 3.2—T4 and T8 lymphocyte counts ($\times 10^9/1$) in 28 haemophiliacs[3]

Subject number	T4	T8	$\log_e(\text{T4}) - \log_e(\text{T8})$
1	0·20	0·17	0·163
2	0·27	0·52	−0·655
3	0·28	0·25	0·113
4	0·37	0·34	0·085
5	0·38	0·14	0·999
6	0·48	0·10	1·569
7	0·49	0·58	−0·169
8	0·56	0·23	0·890
9	0·60	0·24	0·916
10	0·64	0·67	−0·046
11	0·64	0·90	−0·341
12	0·66	0·26	0·932
13	0·70	0·51	0·317
14	0·77	0·18	1·453
15	0·88	0·74	0·173
16	0·88	0·54	0·488
17	0·88	0·76	0·147
18	0·90	0·62	0·373
19	1·02	0·48	0·754
20	1·10	0·58	0·640
21	1·10	0·34	1·174
22	1·18	0·84	0·340
23	1·20	0·63	0·644
24	1·30	0·46	1·039
25	1·40	0·84	0·511
26	1·60	1·20	0·288
27	1·64	0·59	1·022
28	2·40	1·30	0·613

The mean of the values of $\log_e(\text{T4})$ is $-0{\cdot}2896$ and the standard deviation is 0·5921. Thus the standard error of the mean is found as $0{\cdot}5921/\sqrt{28} = 0{\cdot}1119$. The units here are log lymphocyte counts $\times 10^9/1$.

To calculate the 95% confidence interval the appropriate value of $t_{0{\cdot}975}$ with 27 degrees of freedom is 2·052. The 95% confidence interval for the mean $\log_e(\text{T4})$ in the population is then given by

$$-0{\cdot}2896 - (2{\cdot}052 \times 0{\cdot}1119) \quad \text{to} \quad -0{\cdot}2896 + (2{\cdot}052 \times 0{\cdot}1119)$$

that is, from $-0{\cdot}5192$ to $-0{\cdot}0600$.

We can transform this confidence interval on the logarithmic scale back to the original units to get a more meaningful confidence interval. First we transform back the mean of $\log_e(\text{T4})$ to get the geometric mean T4 count. This is given as $\exp(-0{\cdot}2896) = 0{\cdot}75 \times 10^9/1$. (The geometric mean is found as the antilog of the mean of the log values.) In the same way we can transform back the values describing the confidence interval to get a 95% confidence interval for the geometric mean T4 lymphocyte count in the population of haemophiliacs, which is thus given by

$$\exp(-0{\cdot}5192) \quad \text{to} \quad \exp(-0{\cdot}0600)$$

that is, from 0·59 to $0{\cdot}94 \times 10^9/1$.

Two samples

For the case of two samples, only the logarithmic transformation is suitable. For paired or unpaired samples the confidence interval for the difference in the means of the transformed data has to be transformed back. For the log transformation the antilog of the difference in sample means on the transformed scale is an estimate of the ratio of the two population (geometric) means, and the antilogged confidence interval for the difference gives a confidence interval for this ratio. Other transformations do not lead to sensible confidence intervals when transformed back, but a non-parametric approach can be used to calculate a confidence interval for the population difference between medians (see chapter 8).

Worked example

Suppose that we wish to calculate a confidence interval for the difference between the T4 and T8 counts in the population of haemophiliacs using the results given in Table 3.2. Inspection of histograms and plots of these data reveals that the distribution of the differences T4 − T8 is skewed, but after logarithmic transformation the differences $\log_e(\text{T4}) - \log_e(\text{T8})$ have a symmetric near Normal distribution. We can thus apply the method given previously for calculating a confidence interval from paired samples. The method makes use of the fact that the difference between the logarithms of two quantities is exactly the same as the logarithm of their ratio. Thus

$$\log_e(\text{T4}) - \log_e(\text{T8}) = \log_e(\text{T4/T8}).$$

The mean of the differences between the logs of the T4 and T8 counts (shown in the final column of table 3.2) is 0·5154 and the standard deviation is 0·5276. Thus the standard error of the mean is found as $0{\cdot}5276/\sqrt{28} = 0{\cdot}0997$.

To calculate the 95% confidence interval the appropriate value of $t_{0{\cdot}975}$ with 27 degrees of freedom is 2·052. The 95% confidence interval for the difference between the mean values of $\log_e(\text{T4})$ and $\log_e(\text{T8})$ in the population of haemophiliacs is then given by

$$0{\cdot}5154 - (2{\cdot}052 \times 0{\cdot}0997) \quad \text{to} \quad 0{\cdot}5154 + (2{\cdot}052 \times 0{\cdot}0997)$$

that is, from 0·3108 to 0·7200.

The confidence interval for the difference between log counts is not as easy to interpret as a confidence interval relating to the actual counts. We can take antilogs of the above values to get a more useful confidence interval. The antilog of the mean difference between log counts is exp(0·5154) = 1·67. Because of the equivalence of the difference $\log_e(T4) - \log_e(T8)$ and $\log_e(T4/T8)$ this value is an estimate of the geometric mean of the ratio T4/T8 in the population. The antilogs of the values describing the confidence interval are exp(0·3108) = 1·36 and exp(0·7200) = 2·05, and these values provide a 95% confidence interval for the geometric mean ratio of T4 to T8 lymphocyte counts in the population of haemophiliacs.

Note that whereas for a single sample the use of the log transformation still leads to a confidence interval in the original units, for paired samples the confidence interval is in terms of a ratio and has no units.

A confidence interval for the difference in the means of two unpaired samples is derived in much the same way as for paired samples. The log data are used to calculate a confidence interval, using the method for unpaired samples given previously. The antilogs of the difference in the means of the log data and the values describing its confidence interval give the geometric mean ratio and its associated confidence interval.

1 Lentner C, ed. *Geigy scientific tables*. 8th ed. Basle: Geigy, 1982:30–3.
2 Armitage P, Berry G. *Statistical methods in medical research*. 2nd ed. Oxford: Blackwell, 1987:109–12.
3 Ball SE, Hows JM, Worslet AM, *et al*. Seroconversion of human T cell lymphotropic virus III (HTLV-III) in patients with haemophilia: a longitudinal study. *Br Med J* 1985;**290**: 1705–6.

4

Calculating confidence intervals for proportions and their differences

MARTIN J GARDNER, DOUGLAS G ALTMAN

The rationale behind the use of confidence intervals has been discussed in chapters 1 and 2. Confidence intervals for proportions, or differences between two proportions, can be constructed similarly to those for means and their differences described in chapter 3.

Worked examples are given to illustrate each method. The calculations have been carried out to full arithmetical precision, as is recommended practice (see chapter 9), but intermediate steps are shown as rounded results.

Confidence intervals convey only the effects of sampling variation on the estimated proportions and their differences and cannot control for other non-sampling errors such as biases in study design, conduct, or analysis.

Single sample

If p is the observed proportion of subjects with some feature in a sample of size n then the standard error of p is $SE = \sqrt{p(1-p)/n}$. The $100(1-\alpha)\%$ confidence interval for the proportion in the population is given by

$$p - (N_{1-\alpha/2} \times SE) \quad \text{to} \quad p + (N_{1-\alpha/2} \times SE),$$

where $N_{1-\alpha/2}$ is the appropriate value from the standard Normal distribution for the $100(1-\alpha/2)$ percentile. Values of $N_{1-\alpha/2}$ can be found from table 1 in part III, statistical textbooks, or *Geigy Scientific Tables*.[1] Thus, for a 95% confidence interval $N_{1-\alpha/2} = 1{\cdot}96$; this value does not depend on the sample size, as it does for the t distribution.

If p is quoted as a percentage rather than as a proportion then $(1-p)$ should be replaced by $(100-p)$ in the SE calculation.

Effectively the proportion, its standard error, and its confidence interval are multiplied by 100 to convert to percentages.

An exact but more complex method of calculating confidence intervals for population proportions is recommended for small samples and values of p away from $\frac{1}{2}$. The method is described by Armitage and Berry.[2] Also, the exact values for 95% and 99% confidence intervals for $n = 2$ to 100 are given in *Geigy Scientific Tables*.[3]

Worked example

Out of 263 patients giving their views on the use of personal computers in general practice, 81 thought that the privacy of their medical file had been reduced.[4] Thus $p = 81/263 = 0{\cdot}31$ and the standard error of p is

$$SE = \sqrt{\frac{0{\cdot}31 \times (1 - 0{\cdot}31)}{263}} = 0{\cdot}028.$$

The 95% confidence interval for the population value of the proportion of patients thinking their privacy was reduced is then given as

$$0{\cdot}31 - (1{\cdot}96 \times 0{\cdot}028) \quad \text{to} \quad 0{\cdot}31 + (1{\cdot}96 \times 0{\cdot}028)$$

that is, from 0·25 to 0·36.

Two samples: unpaired case

The confidence interval for the difference between two population proportions is constructed round $p_1 - p_2$, the difference between the observed proportions in the two samples of sizes n_1 and n_2. The standard error of $p_1 - p_2$ in this case is

$$SE_{diff} = \sqrt{\frac{p_1(1-p_1)}{n_1} + \frac{p_2(1-p_2)}{n_2}}.$$

The confidence interval for the population difference in proportions is then given by

$$p_1 - p_2 - (N_{1-\alpha/2} \times SE_{diff}) \quad \text{to} \quad p_1 - p_2 + (N_{1-\alpha/2} \times SE_{diff}),$$

where $N_{1-\alpha/2}$ is found as for the single sample case.

If p_1 and p_2 are quoted as percentages rather than proportions then $(1 - p_1)$ and $(1 - p_2)$ should be replaced by $(100 - p_1)$ and $(100 - p_2)$ in the SE_{diff} calculation. Effectively, the difference between proportions, its standard error, and its confidence interval are multiplied by 100 to convert to percentages.

The formulae above should not be used for small samples—for example, fewer than 30 in each group and proportions outside the range 0·1 to 0·9. A continuity correction can be incorporated,[5] as is sometimes done for the χ^2 test of the difference between proportions in a 2×2 table, but is not essential.[6]

Worked example

Response to treatment was assessed among 160 patients randomised to either treatment A or treatment B; the results are shown in table 4.1.

TABLE 4.1—Numbers of patients responding to treatment in groups A and B

	Treatment	
Response	A	B
Improvement	61	45
No improvement	19	35
Total	80	80

The proportions whose condition improved were $p_A = 0{\cdot}76$ and $p_B = 0{\cdot}56$ (61/80 and 45/80) for treatments A and B respectively, which indicates a preferential improvement proportion of 0·20 for treatment A. In terms of percentages 76% of patients on treatment A improved compared with 56% on treatment B, suggesting that an extra 20% of patients would improve if given A rather than B.

The standard error of the difference $p_A - p_B = 0{\cdot}20$ from the formula for the unpaired case is

$$SE_{diff} = \sqrt{\frac{0{\cdot}76 \times 0{\cdot}24}{80} + \frac{0{\cdot}56 \times 0{\cdot}44}{80}} = 0{\cdot}073.$$

The 95% confidence interval for the difference between the two population proportions is then given by

$$0{\cdot}20 - (1{\cdot}96 \times 0{\cdot}073) \quad \text{to} \quad 0{\cdot}20 + (1{\cdot}96 \times 0{\cdot}073)$$

that is, from 0·06 to 0·34. Thus, although the best estimate of the difference in the percentage of patients improving is 20%, the 95% confidence interval ranges from 6% to 34%, showing the imprecision due to the limited sample size.

Two samples: paired case

Suppose that a sample of n subjects has twice been examined for the presence or absence of a particular feature. The data can be presented as in table 4.2.

TABLE 4.2—Classification of subjects by presence or absence of a feature at two times

Feature at time		
1	2	Number of subjects
Present	Present	r
Present	Absent	s
Absent	Present	t
Absent	Absent	u
Total		n

Then the proportions of subjects with the feature on the two occasions are $p_1 = (r+s)/n$ and $p_2 = (r+t)/n$, and the difference between them is $p_1 - p_2 = (s-t)/n$. The standard error of this difference is

$$SE_{diff} = \frac{1}{n} \times \sqrt{s + t - \frac{(s-t)^2}{n}}.$$

The confidence interval for the population value of the difference between proportions is then given as

$$p_1 - p_2 - (N_{1-\alpha/2} \times SE_{diff}) \quad \text{to} \quad p_1 - p_2 + (N_{1-\alpha/2} \times SE_{diff}),$$

where $N_{1-\alpha/2}$ is found as for the single sample case.

If p_1 and p_2 are quoted as percentages rather than proportions then the difference between proportions, its standard error, and its confidence interval are multiplied by 100 to convert to percentages.

This same method is used for matched case-control comparisons as well as studies of repeated observations (see also chapter 6).

An exact method is available and should be used when the numbers in the study are small.[7] The approach is similar to that described in chapter 6 to find the confidence interval for the odds ratio in a matched case-control study. First find A_L and A_U as described in the appropriate section of chapter 6, and then the $100(1-\alpha)\%$ confidence interval for the difference between the population proportions is given as

$$(2A_L - 1) \times \frac{(s+t)}{n} \quad \text{to} \quad (2A_U - 1) \times \frac{(s+t)}{n}.$$

Worked example

Thirty five patients who died in hospital from asthma were individually matched for sex and age with 35 control patients who had been discharged alive from the same hospital in the preceding year.[8] The inadequacy of monitoring of all patients while in hospital was independently assessed; the paired results are given in table 4.3. The proportions with inadequate monitoring were thus $p_1 = 0{\cdot}66$ and $p_2 = 0{\cdot}37$ (23/35 and 13/35) for those who died and survived respectively, which shows a difference of 0·29 in the proportions with inadequate monitoring.

TABLE 4.3—Inadequacy of monitoring in hospital of deaths and survivors among 35 matched pairs of asthma patients[8]

Inadequacy of monitoring		
Deaths	Survivors	Number of pairs
Yes	Yes	10
Yes	No	13
No	Yes	3
No	No	9
Total		35

The standard error of the difference $p_1 - p_2$ is found from the formula for the paired case as

$$SE_{diff} = \frac{1}{35} \times \sqrt{13 + 3 - \frac{(13-3)^2}{35}} = 0{\cdot}104.$$

The 95% confidence interval for the difference between the two population proportions is then given by

$$0{\cdot}29 - (1{\cdot}96 \times 0{\cdot}104) \quad \text{to} \quad 0{\cdot}29 + (1{\cdot}96 \times 0{\cdot}104)$$

that is, from 0·08 to 0·49.

The exact method for this example (see chapter 6 for details) gives a 95% confidence interval of

$$(2 \times 0{\cdot}5435 - 1) \times \frac{16}{35} \quad \text{to} \quad (2 \times 0{\cdot}9595 - 1) \times \frac{16}{35}$$

that is, from 0·04 to 0·42, which shows that the usual approach based on the Normal approximation for large samples is fairly satisfactory in this case.

(For another analysis of these same data see chapter 6.)

Technical note

Although for quantitative data and means there is a direct correspondence between the confidence interval approach and a *t* test of the null hypothesis at the associated level of statistical significance, this is not exactly so for qualitative data and proportions. The reason is related to the use of different estimates of the standard error for the tests of the null hypothesis from those given here for constructing confidence intervals. The lack of direct correspondence is small and should not result in changes of interpretation.

1 Lentner C, ed. *Geigy scientific tables*. 8th ed. Basle: Geigy, 1982:26–9.
2 Armitage P, Berry G. *Statistical methods in medical research*. 2nd ed. Oxford: Blackwell, 1987:117–20.
3 Lentner C, ed. *Geigy scientific tables*. 8th ed. Basle: Geigy, 1982:89–102.
4 Rethans J-J, Hoppener P, Wolfs G, Diederiks J. Do personal computers make doctors less personal? *Br Med J* 1988;**296**:1446–8.
5 Fleiss JL. *Statistical methods for rates and proportions*. 2nd ed. Chichester: Wiley, 1981:29–30.
6 Armitage P, Berry G. *Statistical methods in medical research*. 2nd ed. Oxford: Blackwell, 1987:120.
7 Armitage P, Berry G. *Statistical methods in medical research*. 2nd ed. Oxford: Blackwell, 1987:123.
8 Eason J, Markowe HLJ. Controlled investigation of deaths from asthma in hospitals in the North East Thames region. *Br Med J* 1987; **294**:1255–8.

5

Calculating confidence intervals for regression and correlation

DOUGLAS G ALTMAN, MARTIN J GARDNER

The most common statistical analyses are those that examine one or two groups of individuals with respect to a single variable, and methods of calculating confidence intervals for means or proportions and their differences have been described in chapters 3 and 4. Also common are those analyses that consider the relation between two variables in one group of subjects. We use regression analysis to predict one variable from another, and correlation analysis to see if the values of two variables are associated. The purposes of these two analyses are distinct, and usually one only should be used.

This chapter outlines the calculation of the linear regression equation for predicting one variable from another and shows how to calculate confidence intervals for the population value of the slope and intercept of the line, for the line itself, and for predictions made using the regression equation. It explains how to obtain a confidence interval for the population value of the difference between the slopes of regression lines from two groups of subjects and how to calculate a confidence interval for the vertical distance between two parallel regression lines. The calculations of confidence intervals for Pearson's correlation coefficient and Spearman's rank correlation coefficient are described.

Worked examples are included to illustrate each method. The calculations have been carried out to full arithmetical precision, as is recommended practice (see chapter 9), but intermediate steps are shown as rounded results. Methods of calculating confidence intervals for different aspects of regression and correlation are demonstrated, but the appropriate ones to use depend on the particular problem being studied.

The interpretation of confidence intervals has been discussed in chapters 1 and 2. Confidence intervals convey only the effects of sampling variation on the estimated statistics and cannot control for other errors such as biases in design, conduct, or analysis.

General form of confidence intervals

The basic method for constructing confidence intervals is as described in chapter 2: appendix 2. Each confidence interval is obtained by subtracting from, and adding to, the estimated statistic a multiple of its standard error (SE). The multiple is determined by the theoretical distribution of the statistic: the t distribution for regression, or the Normal distribution for correlation. The multiple is taken as the value that corresponds to including the central $100(1-\alpha)\%$ of the theoretical distribution. So, for example, a 95% confidence interval is described by finding the value that cuts off $2\frac{1}{2}\%$ from each tail of the distribution (see notation list at the end of the book). Tables of the t and Normal distributions are given in part III and are available in most statistics books and *Geigy Scientific Tables*.[1] We denote the relevant value as either $t_{1-\alpha/2}$ or $N_{1-\alpha/2}$. For the t distribution the number of degrees of freedom, which depends on the sample size, must be known.

Regression analysis

For two variables x and y we wish to calculate the regression equation for predicting y from x. We call y the dependent variable and x the independent variable. The equation for the population regression line is

$$y = \mathrm{A} + \mathrm{B}x,$$

where A is the intercept on the vertical y axis (the value of y when $x = 0$) and B is the slope of the line. In standard regression analysis it is assumed that the distribution of the y variable at each value of x is Normal with the same variance, but no assumptions are made about the distribution of the x variable. Sample estimates a (of A) and b (of B) are needed and also the means of the two variables ($\bar{x}$ and $\bar{y}$), the standard deviations of the two variables (s_x and s_y), and the residual standard deviation of y about the regression line (s_{res}). The formulas for deriving a, b, and s_{res} are given in the appendix.

All the following confidence intervals associated with a single regression line use the quantity $t_{1-\alpha/2}$, the appropriate value from the t distribution with $n-2$ degreees of freedom where n is the sample size.

A fitted regression line should be used to make predictions only within the observed range of the x variable. Extrapolation outside this range is unwarranted and may mislead.

It is always advisable to plot the data to see whether a linear relationship between x and y is reasonable. In addition a plot of the "residuals" ($y - y_{fit}$ or "observed minus predicted")—see appendix—is useful to check the distributional assumptions for the y variable.

Illustrative data set

Table 5.1 shows data from a clinical trial of enalaparil versus placebo in diabetic patients.[2] The variables studied are mean arterial blood pressure (mm Hg) and total glycosylated haemoglobin concentration (%). The analyses presented here are illustrative and do not relate directly to the clinical trial. Most of the

TABLE 5.1—Mean arterial blood pressure and total glycosylated haemoglobin concentration in two groups of 10 diabetics on entry to a clinical trial of enalaparil versus placebo[2]

Enalaparil group		Placebo group	
Mean arterial blood pressure (mm Hg) x	Total glycosylated haemoglobin (%) y	Mean arterial blood pressure (mm Hg) x	Total glycosylated haemoglobin (%) y
91	9·8	98	9·5
104	7·4	105	6·7
107	7·9	100	7·0
107	8·3	101	8·6
106	8·3	99	6·7
100	9·0	87	9·5
92	9·7	98	9·0
92	8·8	104	7·6
105	7·6	106	8·5
108	6·9	90	8·6
Means:			
$\bar{x} = 101{\cdot}2$	$\bar{y} = 8{\cdot}37$	$\bar{x} = 98{\cdot}8$	$\bar{y} = 8{\cdot}17$
Standard deviations:			
$s_x = 6{\cdot}941$	$s_y = 0{\cdot}9615$	$s_x = 6{\cdot}161$	$s_y = 1{\cdot}0914$
Standard deviations about the fitted regression lines:			
	$s_{res} = 0{\cdot}5485$		$s_{res} = 0{\cdot}9866$

methods for calculating confidence intervals are demonstrated using only the data from the 10 subjects who received enalaparil.

Single sample

We want to describe the way total glycosylated haemoglobin concentration (TGH) changes with mean arterial blood pressure (MAP). The regression line of total glycosylated haemoglobin concentration on mean arterial blood pressure for the 10 subjects receiving enalaparil is found to be

$$\text{TGH} = 20{\cdot}19 - 0{\cdot}1168 \times \text{MAP}.$$

The estimated slope of the line is negative, indicating lower total glycosylated haemoglobin concentrations for subjects with higher mean arterial blood pressure.

The other quantities needed to obtain the various confidence intervals are shown in table 5.1. The calculations use 95% confidence intervals. For this we need the value of $t_{0{\cdot}975}$ with 8 degrees of freedom, and table 2 in part III shows this to be 2·306.

CONFIDENCE INTERVAL FOR THE SLOPE OF THE REGRESSION LINE

The slope of the sample regression line estimates the mean change in y for a unit change in x. The standard error of the slope, b, is calculated as

$$\text{SE}(b) = \frac{s_{\text{res}}}{s_x\sqrt{n-1}}.$$

The $100(1-\alpha)\%$ confidence interval for the population value of the slope, B, is then given by

$$b - [t_{1-\alpha/2} \times \text{SE}(b)] \quad \text{to} \quad b + [t_{1-\alpha/2} \times \text{SE}(b)].$$

Worked example

The standard error of the slope is

$$\text{SE}(b) = \frac{0{\cdot}5845}{6{\cdot}941 \times \sqrt{9}} = 0{\cdot}026\,34\% \text{ per mm Hg.}$$

The 95% confidence interval for the population value of the slope is thus

$$-0{\cdot}1168 - (2{\cdot}306 \times 0{\cdot}026\,34) \quad \text{to} \quad -0{\cdot}1168 + (2{\cdot}306 \times 0{\cdot}026\,34)$$

that is, from $-0{\cdot}178$ to $-0{\cdot}056\%$ per mm Hg.

CONFIDENCE INTERVAL FOR THE MEAN VALUE OF y FOR A GIVEN VALUE OF x (AND FOR THE REGRESSION LINE)

The estimated mean value of y for any chosen value of x, say x_0, is obtained from the fitted regression line as

$$y_{\text{fit}} = a + bx_0.$$

The standard error of y_{fit} is given by

$$\text{SE}(y_{\text{fit}}) = s_{\text{res}} \times \sqrt{\frac{1}{n} + \frac{(x_0 - \bar{x})^2}{(n-1)s_x^2}}.$$

The $100(1-\alpha)\%$ confidence interval for the population mean value of y at $x = x_0$ is then

$$y_{\text{fit}} - [t_{1-\alpha/2} \times \text{SE}(y_{\text{fit}})] \quad \text{to} \quad y_{\text{fit}} + [t_{1-\alpha/2} \times \text{SE}(y_{\text{fit}})].$$

When this calculation is made for all values of x in the observed range a $100(1-\alpha)\%$ confidence interval for the position of the population regression line is obtained. Because of the last term in the formula for $\text{SE}(y_{\text{fit}})$ the confidence interval becomes wider with increasing distance of x_0 from $\bar{x}$.

Worked example

The confidence interval for the mean total glycosylated haemoglobin concentration can be calculated for any specified value of mean arterial blood pressure. If the mean arterial blood pressure of interest is 100 mm Hg the estimated total glycosylated haemoglobin concentration (y_{fit}) is $20{\cdot}19 - (0{\cdot}1168 \times 100) = 8{\cdot}51\%$. The standard error of this estimated value is

$$\text{SE}(y_{\text{fit}}) = 0{\cdot}5485 \times \sqrt{\frac{1}{10} + \frac{(100 - 101{\cdot}2)^2}{9 \times 6{\cdot}941^2}} = 0{\cdot}1763\%.$$

The 95% confidence interval for the mean total glycosylated haemoglobin concentration for the population of diabetic subjects with a mean arterial blood pressure of 100 mm Hg is thus

$$8{\cdot}51 - (2{\cdot}306 \times 0{\cdot}1763) \quad \text{to} \quad 8{\cdot}51 + (2{\cdot}306 \times 0{\cdot}1763)$$

that is, from 8·10 to 8·92%.

By calculating the 95% confidence interval for the mean total glycosylated haemoglobin concentration for all values of mean arterial blood pressure within the range of observations we get a 95% confidence interval for the population regression line. This is shown in fig 5.1. The confidence interval becomes wider, moving away from the mean mean arterial blood pressure of 101·2 mm Hg.

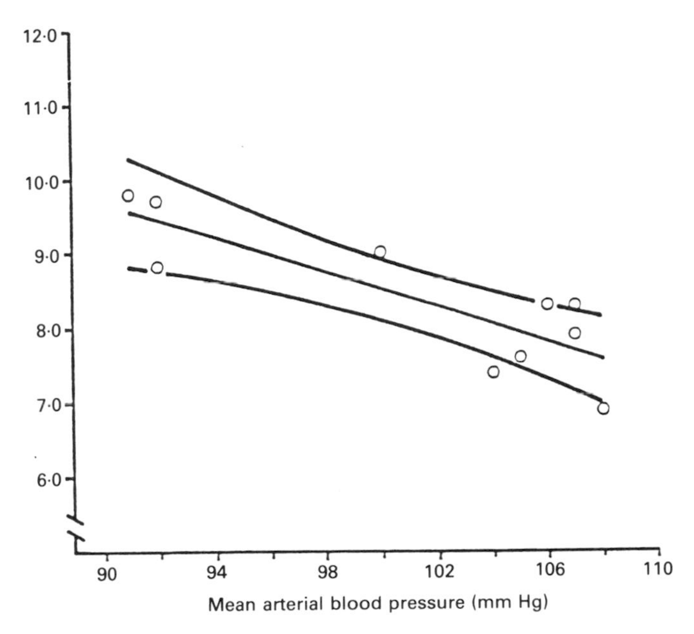

FIG 5.1—Regression line of total glycosylated haemoglobin concentration on mean arterial blood pressure, with 95% confidence interval for the population mean total glycosylated haemoglobin concentration.

CONFIDENCE INTERVAL FOR THE INTERCEPT OF THE REGRESSION LINE

The intercept of the regression line on the y axis is generally of less interest than the slope of the line and does not usually have any obvious interpretation. It can be seen that the intercept is the fitted value of y when x is zero.

Thus a $100(1-\alpha)\%$ confidence interval for the population value of the intercept, A, can be obtained using the formula from the preceding section with $x_0 = 0$ and $y_{\text{fit}} = a$. The standard error of a is given by

$$\text{SE}(a) = s_{\text{res}} \times \sqrt{\frac{1}{n} + \frac{\bar{x}^2}{(n-1)s_x^2}}.$$

The confidence interval is thus given by

$$a - [t_{1-\alpha/2} \times \text{SE}(a)] \quad \text{to} \quad a + [t_{1-\alpha/2} \times \text{SE}(a)].$$

Worked example

The confidence interval for the population value of the intercept is the confidence interval for y_{fit} when $x = 0$, and is calculated as before. In this case the intercept is 20·19%, with a standard error of 2·67%. Thus the 95% confidence interval is from 14·03 to 26·35%. Clearly in this example the intercept, relating to a mean arterial blood pressure of zero, is extrapolated well outside the range of the data and is of no interest in itself.

PREDICTION INTERVAL FOR AN INDIVIDUAL (AND ALL INDIVIDUALS)

It is useful to calculate the uncertainty in y_{fit} as a predictor of y for an individual subject. The range of uncertainty is called a prediction (or tolerance) interval. A prediction interval is wider than the associated confidence interval for the mean value of y because the scatter of data about the regression line is more important. For an individual whose value of x is x_0 the predicted value of y is y_{fit}, given by

$$y_{\text{fit}} = a + bx_0.$$

To calculate the prediction interval we use the estimated standard deviation of individual values of y when x equals x_0 (s_{pred}):

$$s_{\text{pred}} = s_{\text{res}} \times \sqrt{1 + \frac{1}{n} + \frac{(x_0 - \bar{x})^2}{(n-1)s_x^2}}\,.$$

The $100(1-\alpha)\%$ prediction interval is then

$$y_{\text{fit}} - (t_{1-\alpha/2} \times s_{\text{pred}}) \quad \text{to} \quad y_{\text{fit}} + (t_{1-\alpha/2} \times s_{\text{pred}}).$$

When this calculation is made for all values of x in the observed range the estimated prediction interval should include the values of y for $100(1-\alpha)\%$ of subjects in the population.

Worked example

The 95% prediction interval for the total glycosylated haemoglobin concentration of an individual subject with a mean arterial blood pressure of 100 mm Hg is obtained by first calculating s_{pred}:

$$s_{\text{pred}} = 0{\cdot}5485 \times \sqrt{1 + \frac{1}{10} + \frac{(100 - 101{\cdot}2)^2}{9 \times 6{\cdot}941^2}} = 0{\cdot}5761\%.$$

The 95% prediction interval is then given by

$$8{\cdot}51 - (2{\cdot}306 \times 0{\cdot}5761) \quad \text{to} \quad 8{\cdot}51 + (2{\cdot}306 \times 0{\cdot}5761)$$

that is, from 7·18 to 9·84%.

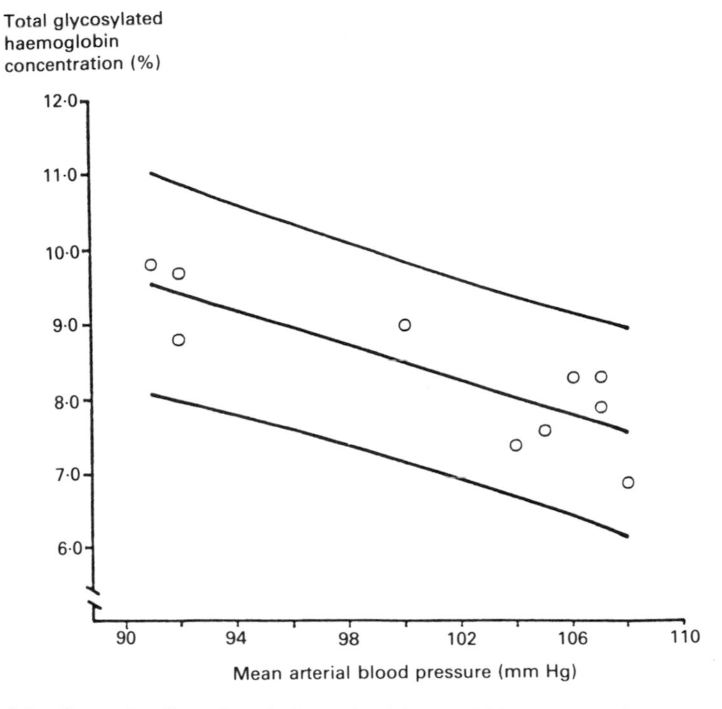

FIG 5.2—Regression line of total glycosylated haemoglobin concentration on mean arterial blood pressure, with 95% prediction interval for an individual total glycosylated haemoglobin concentration.

The contrast with the narrower 95% confidence interval for the mean total glycosylated haemoglobin concentration for a mean arterial blood pressure of 100 mm Hg calculated above is noticeable. The 95% prediction intervals for the range of observed levels of mean arterial blood pressure are shown in fig. 5.2 and again these widen on moving away from the mean arterial blood pressure of 101·2 mm Hg.

Two samples

Regression lines fitted to observations from two independent groups of subjects can be analysed to see if they come from populations with regression lines that are parallel or even coincident.[3]

CONFIDENCE INTERVAL FOR THE DIFFERENCE BETWEEN THE SLOPES OF TWO REGRESSION LINES

If we have fitted regression lines to two different sets of data on the same two variables we can construct a confidence interval for the difference between the population regression slopes using a similar approach to that for a single regression line. The standard error of the difference between the slopes is given by first calculating $s^{\star}_{res}$, the pooled residual standard deviation, as

$$s^{\star}_{res} = \sqrt{\frac{(n_1 - 2)s^2_{res1} + (n_2 - 2)s^2_{res2}}{n_1 + n_2 - 4}}$$

and then

$$SE(b_1 - b_2) = s^{\star}_{res} \times \sqrt{\frac{1}{(n_1 - 1)s^2_{x_1}} + \frac{1}{(n_2 - 1)s^2_{x_2}}},$$

where the suffixes 1 and 2 indicate values derived from the two separate sets of data.

The $100(1 - \alpha)\%$ confidence interval for the population difference between the slopes is now given by

$$b_1 - b_2 - [t_{1-\alpha/2} \times SE(b_1 - b_2)] \quad \text{to} \quad b_1 - b_2 + [t_{1-\alpha/2} \times SE(b_1 - b_2)],$$

where $t_{1-\alpha/2}$ is the appropriate value from the t distribution with $n_1 + n_2 - 4$ degrees of freedom.

Worked example

The regression line for the placebo group from the data in table 5.1 is

$$TGH = 17{\cdot}33 - 0{\cdot}092\,68 \times MAP.$$

The difference between the estimated slopes of the two regression lines is $-0{\cdot}1168 - (-0{\cdot}092\,68) = -0{\cdot}024\,12\%$ per mm Hg. The standard error of this difference is found by first calculating $s^{\star}_{res}$ as

$$s^{\star}_{res} = \sqrt{\frac{(8 \times 0{\cdot}5485^2 + 8 \times 0{\cdot}9866^2)}{16}} = 0{\cdot}7982\% \text{ per mm Hg}$$

and then

$$SE(b_1 - b_2) = 0{\cdot}7982 \times \sqrt{\frac{1}{9 \times 6{\cdot}941^2} + \frac{1}{9 \times 6{\cdot}161^2}}$$

$$= 0{\cdot}057\,74\% \text{ per mm Hg}.$$

From table 2 in part III it is found that the value of $t_{0{\cdot}975}$ with 16 degrees of freedom is 2·120, so the 95% confidence interval for the population difference between the slopes is

$$-0{\cdot}024\,12-(2{\cdot}120\times0{\cdot}057\,74)\quad\text{to}\quad-0{\cdot}024\,12+(2{\cdot}120\times0{\cdot}057\,74)$$

that is, from $-0{\cdot}147$ to $0{\cdot}098\%$ per mm Hg.

Since a zero difference between slopes is near the middle of this confidence interval there is no evidence that the two population regression lines have different slopes. This is not surprising in this example as the subjects were allocated at random to the treatment groups.

CONFIDENCE INTERVAL FOR THE COMMON SLOPE OF TWO PARALLEL REGRESSION LINES

If the chosen confidence interval—for example, 95%—for the difference between population values of the slopes includes zero it is reasonable to fit two parallel regression lines with the same slope and calculate a confidence interval for their common slope.

All the calculation can be done using the results obtained by fitting separate regression lines to the two groups and the standard deviations of the x and y values in the two groups: s_{x_1}, s_{x_2}, s_{y_1}, and s_{y_2}. First define the quantity w as

$$w=(n_1-1)s_{x_1}^2+(n_2-1)s_{x_2}^2.$$

The common slope of the parallel lines (b_{par}) is estimated as

$$b_{\text{par}}=\frac{b_1(n_1-1)s_{x_1}^2+b_2(n_2-1)s_{x_2}^2}{w}.$$

The residual standard deviation of y around the parallel lines (s_{par}) is given by

$$s_{\text{par}}=\sqrt{\frac{(n_1-1)s_{y_1}^2+(n_2-1)s_{y_2}^2-b_{\text{par}}^2\times w}{n_1+n_2-3}}$$

and the standard error of the slope by

$$\text{SE}(b_{\text{par}})=\frac{s_{\text{par}}}{\sqrt{w}}.$$

The $100(1-\alpha)\%$ confidence interval for the population value of the common slope is then

$$b_{\text{par}}-[t_{1-\alpha/2}\times\text{SE}(b_{\text{par}})]\quad\text{to}\quad b_{\text{par}}+[t_{1-\alpha/2}\times\text{SE}(b_{\text{par}})],$$

where $t_{1-\alpha/2}$ is the appropriate value from the t distribution with n_1+n_2-3 degrees of freedom.

Worked example

First calculate the quantity w as

$$w = 9 \times 6{\cdot}941^2 + 9 \times 6{\cdot}161^2$$
$$= 775{\cdot}22.$$

The common slope of the parallel lines is then found as

$$b_{par} = \frac{-0{\cdot}1168 \times 9 \times 6{\cdot}941^2 + (-0{\cdot}092\,68) \times 9 \times 6{\cdot}161^2}{775{\cdot}22}$$
$$= -0{\cdot}1062\% \text{ per mm Hg.}$$

The residual standard deviation of y around the parallel lines is

$$s_{par} = \sqrt{\frac{9 \times 0{\cdot}9615^2 + 9 \times 1{\cdot}0914^2 - (-0{\cdot}1062)^2 \times 775{\cdot}2}{10 + 10 - 3}}$$
$$= 0{\cdot}7786\%$$

and the standard error of the common slope is thus

$$\text{SE}(b_{par}) = \frac{0{\cdot}7786}{\sqrt{775{\cdot}22}} = 0{\cdot}027\,96\% \text{ per mm Hg.}$$

From table 2 in part III the value of $t_{0{\cdot}975}$ with 17 degrees of freedom is 2·110, so that the 95% confidence interval for the population value of b_{par} is therefore

$$-0{\cdot}1062 - (2{\cdot}110 \times 0{\cdot}027\,96) \quad \text{to} \quad -0{\cdot}1062 + (2{\cdot}110 \times 0{\cdot}027\,96)$$

that is, from $-0{\cdot}165$ to $-0{\cdot}047\%$ per mm Hg.

CONFIDENCE INTERVAL FOR THE VERTICAL DISTANCE BETWEEN TWO PARALLEL REGRESSION LINES

The intercepts of the two parallel lines with the y axis are given by

$$\bar{y}_1 - b_{par}\bar{x}_1 \quad \text{and} \quad \bar{y}_2 - b_{par}\bar{x}_2.$$

We are usually more interested in the difference between the intercepts, which is the vertical distance between the parallel lines. This is the same as the difference between the fitted y values for the two groups at the same value of x, and is equivalent to adjusting the observed mean values of y for the mean values of x, a method known as analysis of covariance.[4] The adjusted mean difference (y_{diff}) is calculated as

$$y_{diff} = \bar{y}_1 - \bar{y}_2 - b_{par}(\bar{x}_1 - \bar{x}_2)$$

and the standard error of y_{diff} is

$$SE(y_{diff}) = s_{par} \times \sqrt{\frac{1}{n_1} + \frac{1}{n_2} + \frac{(\bar{x}_1 - \bar{x}_2)^2}{w}}.$$

The $100(1-\alpha)\%$ confidence interval for the population value of y_{diff} is then

$$y_{diff} - [t_{1-\alpha/2} \times SE(y_{diff})] \quad \text{to} \quad y_{diff} + [t_{1-\alpha/2} \times SE(y_{diff})]$$

where $t_{1-\alpha/2}$ is the appropriate value from the t distribution with $n_1 + n_2 - 3$ degrees of freedom.

Worked example

Using the calculated value for the common slope the adjusted difference between the mean total glycosylated haemoglobin concentration in the two

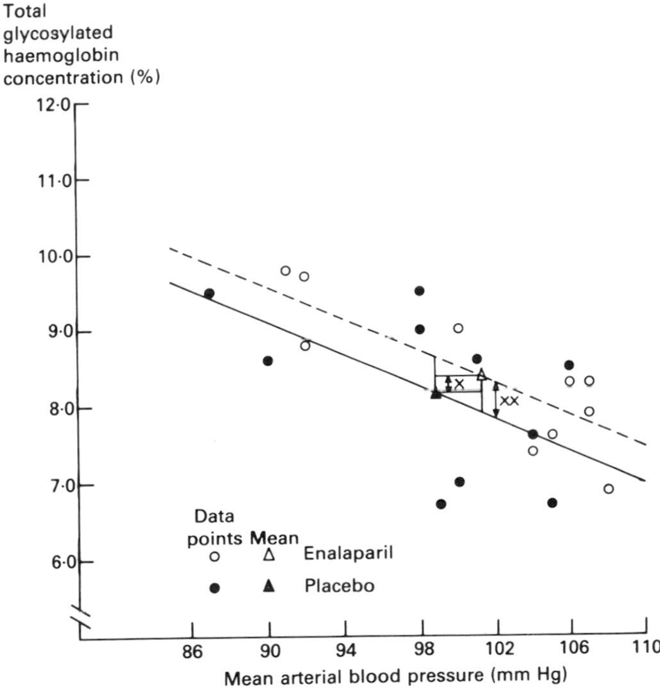

FIG 5.3—Illustration of the calculation of the adjusted difference between mean total glycosylated haemoglobin concentrations in two groups.
Differences between means:

×, observed difference $= \bar{y}_1 - \bar{y}_2 = 0{\cdot}20\%$;
× ×, adjusted difference $= y_{diff} = 0{\cdot}45\%$.

groups is

$$y_{\text{diff}} = (8{\cdot}37 - 8{\cdot}17) - (-0{\cdot}1062) \times (101{\cdot}2 - 98{\cdot}8) = 0{\cdot}4548\%,$$

and its standard error is

$$\text{SE}(y_{\text{diff}}) = 0{\cdot}7786 \times \sqrt{\frac{1}{10} + \frac{1}{10} + \frac{(101{\cdot}2 - 98{\cdot}8)^2}{775{\cdot}22}}$$

$$= 0{\cdot}3546\%.$$

The 95% confidence interval for the population value of y_{diff} is then given by

$$0{\cdot}4548 - (2{\cdot}110 \times 0{\cdot}3546) \quad \text{to} \quad 0{\cdot}4548 + (2{\cdot}110 \times 0{\cdot}3546)$$

that is, from $-0{\cdot}29$ to $1{\cdot}20\%$.

Figure 5.3 illustrates the effect of adjustment.

Extensions

The ideas introduced in this section can be extended to studies with more than two groups and where there are more than two variables by using analysis of covariance and multiple regression[5]. Confidence intervals are constructed in much the same way, and are based on the relevant standard error. These more complex situations are beyond the scope of this book.

Correlation analysis

Pearson's product moment correlation coefficient

The correlation coefficient usually calculated is the product moment correlation coefficient or Pearson's r. This measures the degree of linear "co-relation" between two variables x and y. The formula for calculating r for a sample of observations is given in the appendix.

A confidence interval for the population value of r, assuming that x and y have a joint bivariate Normal distribution, can be constructed by using a transformation of r to a quantity z, which has an approximately Normal distribution. This transformed value, z, is given by

$$z = \tfrac{1}{2} \log_e \frac{1 + r}{1 - r},$$

which for all values of r has a standard error of $1/\sqrt{n - 3}$ where n is the sample size.

For a $100(1 - \alpha)\%$ confidence interval we then calculate the two quantities

$$z_1 = z - \frac{N_{1-\alpha/2}}{\sqrt{n-3}}$$

and

$$z_2 = z + \frac{N_{1-\alpha/2}}{\sqrt{n-3}},$$

where $N_{1-\alpha/2}$ is the appropriate value from the standard Normal distribution for the $100(1 - \alpha/2)$ percentile. This can be found in table 1 in part III.

The values z_1 and z_2 need to be transformed back to the original scale to give a $100(1 - \alpha)\%$ confidence interval for the population correlation coefficient as

$$\frac{e^{2z_1} - 1}{e^{2z_1} + 1} \quad \text{to} \quad \frac{e^{2z_2} - 1}{e^{2z_2} + 1}.$$

Worked example

Table 5.2 shows the basal metabolic rate and total energy expenditure in 24 hours from a study of 13 non-obese women.[6] The data are ranked by increasing basal metabolic rate.

TABLE 5.2—Basal metabolic rate and isotopically measured 24 hour energy expenditure in 13 non-obese women[6]

Basal metabolic rate (MJ/day)	24 hour total energy expenditure (MJ)
4·67	7·05
5·06	6·13
5·31	8·09
5·37	8·08
5·54	7·53
5·65	7·58
5·76	8·40
5·85	7·48
5·86	7·48
5·90	8·11
5·91	7·90
6·19	10·88
6·40	10·15

Pearson's r for these data is 0·7283, and the transformed value z is

$$z = \frac{1}{2}\log_e\frac{1 + 0{\cdot}7283}{1 - 0{\cdot}7283} = 0{\cdot}9251.$$

The values of z_1 and z_2 for a 95% confidence interval are

$$z_1 = 0\cdot9251 - \frac{1\cdot96}{\sqrt{10}} = 0\cdot3053$$

and

$$z_2 = 0\cdot9251 + \frac{1\cdot96}{\sqrt{10}} = 1\cdot545.$$

From these values we derive the 95% confidence interval for the population correlation coefficient as

$$\frac{e^{2\times0\cdot3053} - 1}{e^{2\times0\cdot3053} + 1} \quad \text{to} \quad \frac{e^{2\times1\cdot545} - 1}{e^{2\times1\cdot545} + 1}$$

that is, from 0·296 to 0·913.

Spearman's rank correlation coefficient

If either the distributional assumptions are not met or the relation between x and y is not linear we can use a rank method to assess a more general relation between the values of x and y. To calculate Spearman's rank correlation coefficient (r_s) the values of x and y for the n individuals have to be ranked separately in order of increasing size from 1 to n. Spearman's rank correlation coefficient is then obtained either by using the standard formula for Pearson's product moment correlation coefficient on the ranks of the two variables, or as shown in the appendix using the difference in their two ranks for each individual. The distribution of r_s is similar to that of Pearson's r, so that confidence intervals can be constructed as shown in the previous section.

Appendix: Formulae for regression and correlation analyses

Regression

The slope of the regression line is given by

$$b = \left(\sum xy - n\bar{x}\bar{y}\right) \Big/ \left(\sum x^2 - n\bar{x}^2\right),$$

where $\sum$ represents summation over the n sample points. The intercept is given by

$$a = \bar{y} - b\bar{x}.$$

The difference between the observed and predicted values of y

for an individual with observed values x_o and y_o is $y_o - y_{fit}$, where $y_{fit} = a + bx_o$. The standard deviation of these differences (called "residuals") is thus a measure of how well the line fits the data.

The residual standard deviation of y about the regression line is

$$s_{res} = \sqrt{\frac{\sum (y - y_{fit})^2}{n - 2}}$$

$$= \sqrt{\frac{\sum y^2 - n\bar{y}^2 - b^2(\sum x^2 - n\bar{x}^2)}{n - 2}}$$

$$= \sqrt{\frac{(n-1)(s_y^2 - b^2 s_x^2)}{n - 2}}.$$

Most statistical computer programs give all the necessary quantities to derive confidence intervals, but you may find that the output refers to s_{res} as the "standard error of the estimate."

Correlation

The correlation coefficient (Pearson's r) is estimated by

$$r = \left(\sum xy - n\bar{x}\bar{y}\right) \Big/ \sqrt{\left(\sum x^2 - n\bar{x}^2\right)\left(\sum y^2 - n\bar{y}^2\right)}.$$

Spearman's rank correlation coefficient is given by

$$r_s = 1 - \frac{6 \sum d_i^2}{n^3 - n},$$

where d_i is the difference in the ranks of the two variables for the ith individual. Alternatively, r_s can be obtained by applying the formula for Pearson's r to the ranks of the variables. The calculation of r_s should be modified when there are tied ranks in the data, but the effect is minimal unless there are many tied ranks.

1 Lentner, C, ed. *Geigy scientific tables*. Vol 2. 8th ed. Basle: Ciba-Geigy, 1982:26–33.
2 Marre M, Leblanc H, Suarez L, Guyenne T-T, Ménard J, Passa P. Converting enzyme inhibition and kidney function in normotensive diabetic patients with persistent microalbuminuria. *Br Med J* 1987;**294**:1448–52.
3 Armitage P, Berry G. *Statistical methods in medical research*. 2nd ed. Oxford: Blackwell, 1987:273–95.
4 Armitage P, Berry G. *Statistical methods in medical research*. 2nd ed. Oxford: Blackwell, 1987:282–95.
5 Armitage P, Berry G. *Statistical methods in medical research*. 2nd ed. Oxford: Blackwell, 1987: Sections 9.5 and 10.1.
6 Prentice AM, Black AE, Coward WA, *et al*. High levels of energy expenditure in obese women. *Br Med J* 1986;**292**:983–7.

6

Calculating confidence intervals for relative risks, odds ratios, and standardised ratios and rates

JULIE A MORRIS, MARTIN J GARDNER

The rationale for using estimation and confidence intervals in making inferences from analytical studies has been explained in chapters 1 and 2. Here we present methods for calculating confidence intervals for some common statistics obtained from epidemiological investigations.

The techniques for obtaining confidence intervals for estimates of relative risks and odds ratios are described. These can come either from an incidence study, where, for example, the frequency of a congenital malformation at birth is compared in two defined groups of mothers, or from a case-control study, where a group of patients with the disease of interest (the cases) is compared with another group of people without the disease (the controls).

The methods of obtaining confidence intervals for standardised disease ratios and rates in studies of incidence, prevalence, and mortality are described. Such rates and ratios are commonly calculated to enable appropriate comparisons to be made between study groups after adjustment for confounding factors like age and sex. The most frequently used standardised indices are the standardised incidence ratio (SIR) and the standardised mortality ratio (SMR).

A worked example is included for each method. The calculations have been carried out to full arithmetical precision, as is recommended practice (see chapter 9), although intermediate steps are shown as rounded results. Some of the methods given here are large sample approximations and are not reliable for small studies. Appropriate design principles for these types of study have to be adhered to since confidence intervals convey only the effects of sampling variation on the precision of the estimated statistics and cannot control for other errors such as biases due to the selection of inappropriate controls or in the methods of collecting the data.

Confidence intervals for relative risks and odds ratios

Incidence study

Suppose that the incidence or frequency of some outcome is assessed in two groups of individuals defined by the presence or absence of some characteristic. The data from such a study can be presented as in table 6.1.

TABLE 6.1—Classification of outcome by group characteristic

	Group characteristic	
Outcome	Present	Absent
Yes	A	B
No	C	D
Total	$A+C$	$B+D$

The outcome probabilities in exposed and unexposed individuals are estimated from the study groups by $A/(A+C)$ and $B/(B+D)$ respectively. An estimate, R, of the relative risk (or risk ratio) from exposure is given by the ratio of these proportions, so that

$$R = \frac{A/(A+C)}{B/(B+D)}.$$

Confidence intervals for the population value of R can be constructed through a logarithmic transformation.[1] The standard error of $\log_e R$ is

$$\text{SE}(\log_e R) = \sqrt{\frac{1}{A} - \frac{1}{A+C} + \frac{1}{B} - \frac{1}{B+D}}.$$

A $100(1-\alpha)\%$ confidence interval for R is found by first calculating the two quantities

$$W = \log_e R - [N_{1-\alpha/2} \times \text{SE}(\log_e R)]$$

and

$$X = \log_e R + [N_{1-\alpha/2} \times \text{SE}(\log_e R)],$$

where $N_{1-\alpha/2}$ is the appropriate value from the standard Normal distribution for the $100(1-\alpha/2)$ percentile found in table 1 in part III or in widely available tables.

The confidence interval for the population value of R is then given by exponentiating W and X as

$$e^W \quad \text{to} \quad e^X.$$

Worked example

Susceptibility to rubella in antenatal patients screened in three public health laboratories in England and Wales was studied.[2] The results are shown in table 6.2.

TABLE 6.2—Susceptibility to rubella in Asian and non-Asian antenatal patients[2]

Susceptibility to rubella	Group characteristic	
	Asians	Non-Asians
Yes	161	748
No	2475	34 020
Total	2636	34 768

An estimate of the relative risk of susceptibility to rubella for Asians compared with non-Asians is

$$R = \frac{161/2636}{748/34\,768} = 2{\cdot}84.$$

The standard error of $\log_e R$ is

$$\sqrt{\frac{1}{161} - \frac{1}{2636} + \frac{1}{748} - \frac{1}{34\,768}} = 0{\cdot}0845$$

from which for a 95% confidence interval

$$W = \log_e 2{\cdot}84 - (1{\cdot}96 \times 0{\cdot}0845) = 0{\cdot}8782$$

and

$$X = \log_e 2{\cdot}84 + (1{\cdot}96 \times 0{\cdot}0845) = 1{\cdot}2094.$$

The 95% confidence interval for the population value of R is then given as

$$e^{0{\cdot}8782} \quad \text{to} \quad e^{1{\cdot}2094}$$

that is, from 2·41 to 3·35.

Unmatched case-control study

Suppose that groups of cases and controls are studied to assess exposure to a suspected causal factor. The data can be presented as in table 6.3.

TABLE 6.3—Classification of exposure among cases and controls

	Exposed		
Study group	Yes	No	Total
Cases	a	b	$a+b$
Controls	c	d	$c+d$
Total	$a+c$	$b+d$	$a+b+c+d$

An approximate estimate of the relative risk for the disease associated with exposure to the factor can be obtained from a case-control study through the odds ratio,[3] the relative risk itself not being directly estimable with this study design. The odds ratio (OR) is given as

$$\mathrm{OR} = \frac{ad}{bc}.$$

A confidence interval for the population value of OR can be constructed using several methods which vary in their ease and accuracy. The method described here (sometimes called the logit method) was devised by Woolf[4] and is widely recommended as a satisfactory approximation. The exception to this is when any of the numbers a, b, c, or d is small, when a more accurate but complex procedure should be used if suitable computer facilities are available. Further discussion and comparison of methods can be found in Breslow and Day.[5] The use of the following approach, however, should not in general lead to any misinterpretation of the results.

The logit method uses the Normal approximation to the distribution of the logarithm of the odds ratio ($\log_e \mathrm{OR}$) in which the standard error of $\log_e \mathrm{OR}$ is

$$\mathrm{SE}(\log_e \mathrm{OR}) = \sqrt{\frac{1}{a} + \frac{1}{b} + \frac{1}{c} + \frac{1}{d}}.$$

A $100(1-\alpha)\%$ confidence interval for OR is found by first calculating the two quantities

$$Y = \log_e \text{OR} - [N_{1-\alpha/2} \times \text{SE}(\log_e \text{OR})]$$

and

$$Z = \log_e \text{OR} + [N_{1-\alpha/2} \times \text{SE}(\log_e \text{OR})],$$

where $N_{1-\alpha/2}$ is the appropriate value from the standard Normal distribution for the $100(1 - \alpha/2)$ percentile (see table 1 in part III).

The confidence interval for the population value of OR is then obtained by exponentiating Y and Z to give

$$e^Y \quad \text{to} \quad e^Z.$$

Worked example

ABO non-secretor state was determined for 114 patients with spondyloarthropathies and 334 controls[6] with the results shown in table 6.4.

TABLE 6.4—ABO non-secretor state for 114 patients with spondyloarthropathies and 334 controls[6]

	ABO non-secretor state		
Study group	Yes	No	Total
Cases	54	60	114
Controls	89	245	334
Total	143	305	448

The estimated odds ratio for spondyloarthropathies with ABO non-secretor state is $\text{OR} = (54 \times 245)/(60 \times 89) = 2{\cdot}48$. The standard error of $\log_e \text{OR}$ is

$$\text{SE}(\log_e \text{OR}) = \sqrt{\frac{1}{54} + \frac{1}{60} + \frac{1}{89} + \frac{1}{245}} = 0{\cdot}2247.$$

For a 95% confidence interval

$$Y = \log_e 2{\cdot}48 - (1{\cdot}96 \times 0{\cdot}2247) = 0{\cdot}4678$$

and

$$Z = \log_e 2{\cdot}48 + (1{\cdot}96 \times 0{\cdot}2247) = 1{\cdot}3487.$$

The 95% confidence interval for the population value of the odds ratio for spondyloarthropathies with ABO non-secretor state is then given as

$$e^{0{\cdot}4678} \quad \text{to} \quad e^{1{\cdot}3487}$$

that is, from 1·59 to 3·85.

More than two levels of exposure

If there are more than two levels of exposure one can be chosen as a baseline with which each of the others is compared. Odds ratios and their associated confidence intervals are then calculated for each comparison in the same way as shown previously for only two levels of exposure.

A series of unmatched case-control studies

A combined estimate is sometimes required when independent estimates of the same odds ratio are available from each of K sets of data—for example, in a stratified analysis to control for confounding variables. A common approach is to use the Mantel–Haenszel pooled estimate of the odds ratio ($\mathrm{OR}_{\mathrm{M-H}}$) which is given by

$$\mathrm{OR}_{\mathrm{M-H}} = \sum \frac{a_i d_i}{n_i} \Big/ \sum \frac{b_i c_i}{n_i},$$

where a_i, b_i, c_i, and d_i are the frequencies in the ith 2×2 table, $n_i = a_i + b_i + c_i + d_i$, and the summation $\sum$ is over $i = 1$ to K for the K tables.[7] A method of calculating confidence intervals for this estimate is described in Armitage and Berry.[8] It is more complex, however, than the following approach which extends the single case-control technique shown earlier. For samples of reasonable size both methods will usually give similar values for the combined odds ratio and confidence interval.

An alternative is to use the logit method to give a pooled estimate of the odds ratio (OR_L) and then derive a confidence interval for the odds ratio in a similar way to that for a single 2×2 table. The logit combined estimate (OR_L) is defined by

$$\log_e \mathrm{OR}_\mathrm{L} = \sum w_i \log_e \mathrm{OR}_i \Big/ \sum w_i,$$

where $\mathrm{OR}_i = a_i d_i / b_i c_i$ is the odds ratio in the ith table and

$$w_i = \frac{1}{1/a_i + 1/b_i + 1/c_i + 1/d_i}.$$

The standard error of $\log_e(\mathrm{OR}_\mathrm{L})$ is given by

$$\mathrm{SE}(\log_e \mathrm{OR}_\mathrm{L}) = \frac{1}{\sqrt{w}},$$

where $w = \sum w_i$.

A 100(1 − α)% confidence interval for OR_L can then be found by calculating

$$M = \log_e OR_L - [N_{1-\alpha/2} \times SE(\log_e OR_L)]$$

and

$$N = \log_e OR_L + [N_{1-\alpha/2} \times SE(\log_e OR_L)],$$

where $N_{1-\alpha/2}$ is the appropriate value from the standard Normal distribution for the 100(1 − $\alpha/2$) percentile (see table 1 in part III).

The confidence interval for the population value of OR_L is given by exponentiating M and N as

$$e^M \quad \text{to} \quad e^N.$$

It is important to mention that, before combining independent estimates, there should be some reassurance that the separate odds ratios do not vary markedly, apart from sampling error.[8 9] Further discussion of methods and a worked example are given in Breslow and Day.[9] The logit method is unsuitable if any of the numbers a_i, b_i, c_i, or d_i is small. This will happen, for example, with increasing stratification, and in such cases a more complex exact method is available.[9]

Worked example

A number of studies examining the relationship of passive smoking exposure to lung cancer were reviewed by Wald *et al.*[10] The findings, which are shown in table 6.5, included those from four studies of women in the USA. They are used below to illustrate the combination of results from independent sets of data.

TABLE 6.5—Exposure to passive smoking among female lung cancer cases and controls in four studies[10]

	Lung cancer cases		Controls		
Study	Exposed (a)	Unexposed (b)	Exposed (c)	Unexposed (d)	Odds ratio
1	14	8	61	72	2·07
2	33	8	164	32	0·80
3	13	11	15	10	0·79
4	91	43	254	148	1·23

The logit combined estimate of the odds ratio (OR_L) over the four studies is found to be 1·19. The standard error of $\log_e(OR_L)$ is 0·1693.

For a 95% confidence interval

$$M = \log_e 1{\cdot}19 - (1{\cdot}96 \times 0{\cdot}1693) = -0{\cdot}1579$$

and

$$N = \log_e 1{\cdot}19 + (1{\cdot}96 \times 0{\cdot}1693) = 0{\cdot}5058.$$

The 95% confidence interval for the population value of the odds ratio of lung cancer associated with passive smoking exposure is then given by

$$e^{-0{\cdot}1579} \quad \text{to} \quad e^{0{\cdot}5058}$$

that is, from 0·85 to 1·66.

If, alternatively, the Mantel–Haenszel method is used on these data the combined estimate of the odds ratio is found to be also 1·19, with a 95% confidence interval of 0·86 to 1·66—virtually the same result.

Matched case-control study

If each of n cases of a disease is matched to one control to form n pairs and each individual's exposure to a suspected causal factor is recorded the data can be presented as in table 6.6.

For this type of study an approximate estimate (in fact the Mantel–Haenszel estimate) of the relative risk of the disease associated with exposure is again given by the odds ratio which is now calculated as

$$\mathrm{OR} = \frac{s}{t}.$$

An exact $100(1 - \alpha)\%$ confidence interval for the population value of OR is found by first determining a confidence interval for s (the number of case-control pairs with only the case exposed).[11] Conditional on the sum of the numbers of "discordant" pairs $(s + t)$

TABLE 6.6—Classification of matched case-control pairs by exposure

Exposure status		
Case	Control	Number of pairs
Yes	Yes	r
Yes	No	s
No	Yes	t
No	No	u
Total		n

the number s can be considered as a Binomial variable with sample size $s + t$ and proportion $s/(s + t)$.

The $100(1 - \alpha)\%$ confidence interval for the population value of the Binomial proportion can be obtained from tables based on the Binomial distribution (for example, ref 12). If this confidence interval is denoted by A_L to A_U the $100(1 - \alpha)\%$ confidence interval for the population value of OR is then given by

$$\frac{A_L}{1 - A_L} \text{ to } \frac{A_U}{1 - A_U}.$$

Worked example

Thirty five patients who died in hospital from asthma were individually matched for sex and age with 35 control subjects who had been discharged alive from the same hospital in the preceding year.[13] The inadequacy of monitoring of all patients while in hospital was independently assessed; the paired results are given in table 6.7.

TABLE 6.7—Inadequacy of monitoring in hospital of deaths and survivors among 35 matched pairs of asthma patients[13]

Inadequacy of monitoring		
Deaths	Survivors	Number of pairs
Yes	Yes	10
Yes	No	13
No	Yes	3
No	No	9
Total		35

The estimated odds ratio of dying in hospital associated with inadequate monitoring is $OR = 13/3 = 4{\cdot}33$.

From the appropriate table for the Binomial distribution with sample size $= 13 + 3 = 16$, proportion $= 13/(13 + 3) = 0{\cdot}81$, and $\alpha = 0{\cdot}05$ the 95% confidence interval for the population value of the Binomial proportion is found to be $A_L = 0{\cdot}5435$ to $A_U = 0{\cdot}9595$. The 95% confidence interval for the population value of the odds ratio is thus

$$\frac{0{\cdot}5435}{1 - 0{\cdot}5435} \text{ to } \frac{0{\cdot}9595}{1 - 0{\cdot}9595}$$

that is, from 1·19 to 23·69.

Matched case-control study with 1 : *M matching*

Sometimes each case is matched with more than one control. The odds ratio is then given by the Mantel–Haenszel estimate as

$$OR_{M\text{-}H} = \sum [(M - i + 1) \times n_{1,i-1}] \Big/ \sum (i \times n_{0,i}),$$

where M is the number of matched controls for each case, $n_{1,i}$ is the number of matched sets in which the case and i controls are exposed, $n_{0,i}$ is the number of sets in which the case is unexposed and i controls are exposed, and the summation is from 1 to M.

A confidence interval for the population value of $OR_{M\text{-}H}$ can be derived by one of the methods in Breslow and Day.[14] These authors also explain the calculation of a confidence interval for the odds ratio estimated from a study with a variable number of matched controls for each case.[15]

Confidence intervals for standardised ratios and rates

Standardised ratios

If O is the observed number of incident cases (or deaths) in a study group and E the expected number based on, for example, the age specific disease incidence (or mortality) rates in a reference or standard population the standardised incidence ratio (SIR) or standardised mortality ratio (SMR) is O/E. This is usually called the indirect method of standardisation. The expected number is calculated as

$$E = \sum n_i R_i,$$

where n_i is the number of individuals in age group i of the study group, R_i is the death rate in age group i of the reference population, and $\sum$ denotes summation over all age groups.

The $100(1 - \alpha)\%$ confidence interval for the population value of O/E can be found by first regarding O as a Poisson variable and findings its related confidence interval.[16] This is derived from tables based on the Poisson distribution, such as table 3 in part III. Denote this confidence interval by O_L to O_U.

The $100(1 - \alpha)\%$ confidence interval for the population value of O/E is then given by

$$\frac{O_L}{E} \text{ to } \frac{O_U}{E}.$$

Worked example

Roman *et al* observed 64 cases of leukaemia in children under the age of 15 years in the West Berkshire Health Authority area during 1972–85.[17] They calculated that 45·6 cases would be expected in the area at the age specific leukaemia registration rates by calendar year for this period in England and Wales (the standard population). Using $O = 64$ and $E = 45{\cdot}6$ the standardised incidence ratio (SIR) is $64/45{\cdot}6 = 1{\cdot}40$. Values of $O_L = 49{\cdot}3$ and $O_U = 81{\cdot}7$ are found from the appropriate table based on the Poisson distribution when $O = 64$ and $\alpha = 0{\cdot}05$ (see table 3 in part III).

The 95% confidence interval for the population value of the standardised incidence ratio is

$$\frac{49{\cdot}3}{45{\cdot}6} \quad \text{to} \quad \frac{81{\cdot}7}{45{\cdot}6}$$

that is, from 1·08 to 1·79.

Sometimes the standardised incidence ratio (or standardised mortality ratio) is multiplied by 100 and then the same must be done to the figures describing the confidence interval.

Ratio of two standardised ratios

Let O_1 and O_2 be the observed numbers of cases (deaths) in two study groups and E_1 and E_2 the two expected numbers. It is sometimes appropriate to calculate the ratio of the two standardised incidence ratios (standardised mortality ratios) O_1/E_1 and O_2/E_2 and find a confidence interval for this ratio. Although there are known limitations to this procedure if the age specific incidence ratios within each group to the standard are not similar, these are not serious in usual applications.[18]

Again O_1 and O_2 can be regarded as Poisson variables and a confidence interval for the ratio O_1/O_2 is obtained as described by Ederer and Mantel.[19] The procedure recognises that conditional on the total of $O_1 + O_2$ the number O_1 can be considered as a Binomial variable with sample size $O_1 + O_2$ and proportion $O_1/(O_1 + O_2)$. The $100(1 - \alpha)$% confidence interval for the population value of the Binomial proportion can be obtained from tables based on the Binomial distribution (for example, ref 12). Denote this confidence interval by A_L to A_U. The $100(1 - \alpha)$% confidence interval for the population value of O_1/O_2 can now be found as

$$B_L = \frac{A_L}{1 - A_L} \quad \text{to} \quad B_U = \frac{A_U}{1 - A_U}.$$

The $100(1-\alpha)\%$ confidence interval for the population value of the ratio of the two standardised incidence ratios (standardised mortality ratios) is then given by

$$B_L \times \frac{E_2}{E_1} \quad \text{to} \quad B_U \times \frac{E_2}{E_1}.$$

Worked example

Roman *et al* published figures for childhood leukaemia during 1972–85 in Basingstoke and North Hampshire Health Authority which gave $O=25$ and $E=23{\cdot}7$ and a standardised incidence ratio of 1·05.[17] To compare the figures for West Berkshire from the previous example with those from Basingstoke and North Hampshire let $O_1=64$, $E_1=45{\cdot}6$, $O_2=25$, and $E_2=23{\cdot}7$.

The ratio of the two standardised incidence ratios is given by $(64/45{\cdot}6)/(25/23{\cdot}7)=1{\cdot}40/1{\cdot}05=1{\cdot}33$. From the appropriate table for the Binomial distribution with sample size $=64+25=89$, proportion $=64/(64+25)=0{\cdot}72$, and $\alpha=0{\cdot}05$, the 95% confidence interval for the population value of the Binomial proportion is found to be $A_L=0{\cdot}6138$ to $A_U=0{\cdot}8093$. The 95% confidence interval for O_1/O_2 is thus $0{\cdot}6138/(1-0{\cdot}6138)$ to $0{\cdot}8093/(1-0{\cdot}8093)$—that is, from 1·59 to 4·24.

The 95% confidence interval for the population value of the ratio of the two standardised incidence ratios is then given by

$$1{\cdot}59 \times \frac{23{\cdot}7}{45{\cdot}6} \quad \text{to} \quad 4{\cdot}24 \times \frac{23{\cdot}7}{45{\cdot}6}$$

that is, from 0·83 to 2·21.

Standardised rates

If a rate rather than a ratio is required the standardised rate (SR) in a study group is given by

$$\mathrm{SR} = \sum N_i r_i \Big/ \sum N_i,$$

where N_i is the number of individuals in age group i of the reference population, r_i is the disease rate in age group i of the study group, and Σ indicates summation over all age groups. This is usually known as the direct method of standardisation. If n_i is the number of individuals in age group i of the study group the approximate standard error of SR is

$$\mathrm{SE(SR)} = \sqrt{\sum (N_i^2 r_i / n_i)} \Big/ \sum N_i,$$

assuming that the rates r_i are small.[20]

The $100(1-\alpha)\%$ confidence interval for the population value of SR is then given by

$$\mathrm{SR} - [N_{1-\alpha/2} \times \mathrm{SE(SR)}] \quad \text{to} \quad \mathrm{SR} + [N_{1-\alpha/2} \times \mathrm{SE(SR)}],$$

where $N_{1-\alpha/2}$ is the appropriate value from the Normal distribution for the $100(1-\alpha/2)$ percentile (see table 1 in part III).

Note that if the rates r_i are given as rates per 10^m (for example, $m = 3$ gives a rate per 1000), rather than as proportions, then the standardised rate (SR) is also a rate per 10^m and SE(SR) as given above needs to be multiplied by $\sqrt{10^m}$.

Worked example

The observations presented in table 6.8 were made in a study of the radiological prevalence of Paget's disease of bone in British male migrants to Australia.[21]

The standardised prevalence rate (SR) is 5·7 per 100 with SE(SR) = 1·17 per 100.

TABLE 6.8—Paget's disease of bone in British male migrants to Australia, by age group[21]

Age (years)	Study group			Standard population (N_i)[22]
	Cases	n_i	Rate per 100 (r_i)	
55–64	4	96	4·2	2773
65–74	13	237	5·5	2556
75–84	8	105	7·6	1113
⩾85	7	32	21·9	184
Totals	32	470	6·8	6626

The 95% confidence interval for the population value of the standardised prevalence rate is then given by

$$5{\cdot}7 - (1{\cdot}96 \times 1{\cdot}17) \quad \text{to} \quad 5{\cdot}7 + (1{\cdot}96 \times 1{\cdot}17)$$

that is, from 3·5 to 8·0 per 100.

Comment

Many of the methods described here and others are also given with examples by Rothman.[23]

1 Katz D, Baptista J, Azen SP, Pike MC. Obtaining confidence intervals for the risk ratio in cohort studies. *Biometrics* 1978;**34**:469–74.
2 Miller E, Nicoll A, Rousseau SA, *et al.* Congenital rubella in babies of south Asian women in England and Wales: an excess and its causes. *Br Med J* 1987;**294**:737–9.
3 Armitage P, Berry G. *Statistical methods in medical research.* 2nd ed. Oxford: Blackwell, 1987:458.
4 Woolf B. On estimating the relationship between blood group and disease. *Ann Hum Genet* 1955;**19**:251–3.
5 Breslow NE, Day NE. *Statistical methods in cancer research.* Vol 1. *The analysis of case-control studies.* Lyons: International Agency for Research on Cancer, 1980: sections 4.2 and 4.3.
6 Shinebaum R, Blackwell CC, Forster PJG, Hurst NP, Weir DM, Nuki G. Non-secretion of ABO blood group antigens as a host susceptibility factor in the spondyloarthropathies. *Br Med J* 1987;**294**:208–10.
7 Breslow NE, Day NE. *Statistical methods in cancer research*: Vol 1. *The analysis of case-control studies.* Lyons: International Agency for Research on Cancer, 1980:140.
8 Armitage P, Berry G. *Statistical methods in medical research.* 2nd ed. Oxford: Blackwell, 1987:461.
9 Breslow NE, Day NE. *Statistical methods in cancer research*: Vol 1. *The analysis of case-control studies.* Lyons: International Agency for Research on Cancer, 1980: section 4.4.
10 Wald NJ, Nanchahal K, Thompson SG, Cuckle HS. Does breathing other people's tobacco smoke cause lung cancer? *Br Med J* 1986;**293**:1217–22.
11 Armitage P, Berry G. *Statistical methods in medical research.* 2nd ed. Oxford: Blackwell, 1987:461,462.
12 Lentner C, ed. *Geigy scientific tables.* Vol 2. 8th ed. Basle: Ciba-Geigy, 1982:89–102.
13 Eason J, Markowe HLJ. Controlled investigation of deaths from asthma in hospitals in the North East Thames region. *Br Med J* 1987;**294**:1255–8.
14 Breslow NE, Day NE. *Statistical methods in cancer research.* Vol. 1. *The analysis of case-control studies.* Lyons: International Agency for Research on Cancer, 1980: section 5.3.
15 Breslow NE, Day NE. *Statistical methods in cancer research.* Vol. 1. *The analysis of case-control studies.* Lyons: International Agency for Research on Cancer, 1980: section 5.4.
16 Breslow NE, Day NE. *Statistical methods in cancer research.* Vol 2. *The design and analysis of cohort studies.* Oxford: University Press, 1988:69.
17 Roman E, Beral V, Carpenter L, *et al.* Childhood leukaemia in the West Berkshire and the Basingstoke and North Hampshire District Health Authorities in relation to nuclear establishments in the vicinity. *Br Med J* 1987;**294**:597–602.
18 Breslow NE, Day NE. *Statistical methods in cancer research.* Vol 2. *The design and analysis of cohort studies.* Oxford: University Press, 1988:93.
19 Ederer F, Mantel N. Confidence limits on the ratio of two Poisson variables. *Am J Epidemiol* 1974;**100**:165–7.
20 Armitage P, Berry G. *Statistical methods in medical research.* 2nd ed. Oxford: Blackwell, 1987:401.
21 Gardner MJ, Guyer PB, Barker DJP. Radiological prevalence of Paget's disease of bone in British migrants to Australia. *Br Med J* 1978;**i**:1655–7.
22 Barker DJP, Clough PWL, Guyer PB, Gardner MJ. Paget's disease of bone in 14 British towns. *Br Med J* 1977;**i**:1181–3.
23 Rothman KJ. Modern epidemiology. Boston: Little, Brown, 1986; chapters 11–13.

7

Calculating confidence intervals for survival time analyses

DAVID MACHIN, MARTIN J GARDNER

It is common in follow up studies to be concerned with the survival time between entry to the study and a subsequent event. The event may be death in a study of cancer, the disappearance of pain in a study comparing different steroids in arthritis, or the return of ovulation after stopping a long acting method of contraception. These studies often generate some so called "censored" observations of survival time. Such an observation would occur, for example, on any patient who is still alive at the time of analysis in a trial where death is the end point. In this case the time from allocation to treatment to the latest follow up visit would be the patient's censored survival time.

The Kaplan–Meier product limit technique is the recognised approach for calculating survival curves in such studies.[1] An outline of this method is given here with details of how to calculate a confidence interval for the population value of the survival proportion at any time during the follow up. The situations of a single group of patients and of the difference in survival proportions between two groups are considered. For the latter case confidence interval calculations are also described for the hazard ratio between groups—for example, the relative death rate, relapse rate, etc.

A worked example is included for each method. The calculations have been carried out to full arithmetical precision, as is recommended practice (see chapter 9), although intermediate steps are shown as rounded results. The rationale behind the use of confidence intervals has been described in chapters 1 and 2. Confidence intervals convey only the effects of sampling variation on the precision of the estimated statistics and cannot control for any non-sampling errors such as bias in the selection of patients or in losses to follow up.

Survival proportions and their differences

Single sample

Suppose that the survival times after entry to the study (ordered by increasing duration) of a group of n subjects are $t_1, t_2, t_3, \ldots t_n$. The proportion of subjects (p) surviving beyond any follow up time (t) is estimated by the Kaplan–Meier technique as

$$p = \prod \frac{r_i - d_i}{r_i},$$

where r_i is the number of subjects alive just before time t_i (the ith ordered survival time), d_i denotes the number who died at time t_i, and $\prod$ indicates multiplication over each time a death occurs up to and including time t.

The standard error (SE) of p is given by

$$\text{SE} = \sqrt{\frac{p(1-p)}{n'}},$$

where n' is the "effective" sample size at time t. When there are no censored survival times n' will be equal to n, the total number of subjects in the study group. When censored observations are present the effective sample size can be calculated as

$$n' = \frac{r_i - d_i}{p}$$

each time a death occurs.[2] One alternative, which is also simple to use, is to find n' as[3]

$$n' = n - \text{number of subjects lost to follow up before time } t.$$

The $100(1 - \alpha)\%$ confidence interval for the population value of the survival proportion p at time t is then calculated as

$$p - (N_{1-\alpha/2} \times \text{SE}) \quad \text{to} \quad p + (N_{1-\alpha/2} \times \text{SE}),$$

where $N_{1-\alpha/2}$ is the appropriate value from the standard Normal distribution for the $100(1 - \alpha/2)$ percentile. This is widely available in tables—for example, table 1 in part III. Thus for a 95% confidence interval $\alpha = 0{\cdot}05$ and $N_{1-\alpha/2} = 1{\cdot}96$.

The times at which to estimate survival proportions and their confidence intervals should be determined in advance of the results. They can be chosen according to practical convention—for example, the five year survival proportions which are often quoted in cancer studies—or according to previous similar studies. The formulae for confidence intervals given in this chapter are not reliable for small sample sizes and for p close to 0 or 1. If n' is less than about 10 or p outside the range 0·1 to 0·9 the confidence intervals should be interpreted with caution.

Worked example

Consider the survival experience of the 25 patients randomly assigned to receive γ linolenic acid for the treatment of colorectal cancer of Dukes's stage C.[4] The ordered survival times (t), the calculated survival propor-

TABLE 7.1—Survival data by month for 49 patients with Dukes's C colorectal cancer randomly assigned to receive either γ linolenic acid or control treatment[4]

Group treated with γ linolenic acid				Controls			
Case No	Survival time* (months) t	Survival proportion** p	Effective sample size† n'	Case No	Survival time* (months) t	Survival proportion** p	Effective sample size† n'
1	1+	1	25	26	3+	1	24
2	5+	1	24	27	6		
3	6	0.9130	23	28	6	0·8261	23
4	6			29	6		
5	9+	"	"	30	6		
6	10			31	8	0·7391	"
7	10	0·8217	22	32	8		
8	10+			33	12		
9	12			34	12	0·6522	"
10	12			35	12+		
11	12	0·6284	21	36	15+	"	22
12	12			37	16+	"	21
13	12+			38	18+	"	20
14	13+	"	20	39	18+		
15	15+	"	19	40	20	0·5870	18
16	16+	"	18	41	22+	"	"
17	20+	"	17	42	24	0·5136	17
18	24	0·5498	16	43	28+		
19	24+			44	28+	"	"
20	27+	"	15	45	28+		
21	32	0·4399	14	46	30	0·3852	14
22	34+	"	"	47	30+		
23	36+	"	13	48	33+	"	13
24	36+			49	42	0	12
25	44+	"	11				

* Survival times are shown in each group by month to either death or to censoring. Figures with plus signs show that patient follow up was censored.
** Figures need not be recalculated except when a death occurs.
† Figures are calculated using method from ref 3 (see text) and need not be recalculated except when a loss to follow up occurs.

tions (p), and the effective sample sizes (n'), according to the method of ref 3, are shown in table 7.1.

The data come from a comparative trial, but it may be of interest to quote the two year survival proportion and its confidence interval for the group receiving γ linolenic acid. The survival proportion to any follow up time is taken from the entries in the table for that time if available or otherwise for the time immediately preceding. Thus for two years the survival proportion is $p = 0{\cdot}5498$ and the effective sample size is $n' = 16$.

The standard error of this survival proportion is

$$SE = \sqrt{\frac{0{\cdot}5498 \times (1 - 0{\cdot}5498)}{16}} = 0{\cdot}1244.$$

The 95% confidence interval for the population value of the survival proportion is then given by

$$0{\cdot}5498 - (1{\cdot}96 \times 0{\cdot}1244) \quad \text{to} \quad 0{\cdot}5498 + (1{\cdot}96 \times 0{\cdot}1244)$$

that is, from 0·31 to 0·79.

The estimated percentage of survivors to two years is thus 55% with a 95% confidence interval of 31% to 79%.

Two samples

The difference between survival proportions at any time t in two study groups of sample sizes n_1 and n_2 is measured by $p_1 - p_2$, where p_1 and p_2 are the survival proportions at time t in groups 1 and 2 respectively.

The standard error of $p_1 - p_2$ is

$$SE_{diff} = \sqrt{\frac{p_1(1-p_1)}{n_1'} + \frac{p_2(1-p_2)}{n_2'}},$$

where n_1' and n_2' are the effective sample sizes at time t in each group.

The $100(1-\alpha)$% confidence interval for the population value of $p_1 - p_2$ is

$$p_1 - p_2 - (N_{1-\alpha/2} \times SE_{diff}) \quad \text{to} \quad p_1 - p_2 + (N_{1-\alpha/2} \times SE_{diff}),$$

where $N_{1-\alpha/2}$ is found as for a single sample.

Worked example

The survival experience of the patients receiving γ linolenic acid and the controls can be compared from the results given in table 7.1. At two years, for example, $p_1 = 0{\cdot}5498$ and $p_2 = 0{\cdot}5136$ with $n_1' = 16$ and $n_2' = 17$. The estimated difference in two year survival proportions is thus 0·5498 − 0·5136 = 0·0363.

The standard error of this difference in survival proportions is

$$\sqrt{\frac{0{\cdot}5498\times(1-0{\cdot}5498)}{16}+\frac{0{\cdot}5136\times(1-0{\cdot}5136)}{17}}=0{\cdot}1737.$$

The 95% confidence interval for the population value of the difference in two year survival proportions is then given by

$$0{\cdot}0363-(1{\cdot}96\times 0{\cdot}1737) \quad \text{to} \quad 0{\cdot}0363+(1{\cdot}96\times 0{\cdot}1737)$$

that is, from −0·30 to 0·38.

Thus the study estimate of the increased survival proportion at two years for the patients given γ linolenic acid compared with the control group is only about 4%. Moreover, the imprecision in the estimate from this small study is indicated by the 95% confidence interval ranging from − 30% to +38%.

The hazard ratio

The ratio of failure—for example, death or relapse—rates in a follow up study of two groups is termed the "hazard ratio" and is a common measure of the relative effect of treatment, exposure, etc. If O_1 and O_2 are the observed numbers of deaths at time t in the two groups then the expected numbers of deaths (E_1 and E_2) assuming an equal risk of dying at each time in both groups may be calculated as

$$E_1=\sum\frac{r_{1i}d_i}{r_i} \quad \text{and} \quad E_2=\sum\frac{r_{2i}d_i}{r_i},$$

where r_{1i} and r_{2i} are the numbers of subjects alive and not censored in groups 1 and 2 just before time t_i with $r_i = r_{1i} + r_{2i}$; $d_i = d_{1i} + d_{2i}$ is the number who died at time t_i in the two groups combined; and $\sum$ indicates addition over each time of death up to and including t.

To obtain a $100(1-\alpha)$% confidence interval for the population value of the hazard ratio first calculate the two quantities

$$X=\frac{O_1-E_1}{V} \quad \text{and} \quad Y=\frac{N_{1-\alpha/2}}{\sqrt{V}},$$

where

$$V=\sum\frac{r_{1i}r_{2i}d_i(r_i-d_i)}{r_i^2(r_i-1)}$$

and $N_{1-\alpha/2}$ is the appropriate value from the standard Normal distribution for the $100(1-\alpha/2)$ percentile (see table 1 in part III). Thus for a 95% confidence interval $\alpha = 0{\cdot}05$ and $N_{1-\alpha/2} = 1{\cdot}96$.

The hazard ratio can then be estimated[5] by e^X and the confidence interval for the hazard ratio by

$$e^{X-Y} \quad \text{to} \quad e^{X+Y}.$$

The hazard ratio is more usually calculated as $(O_1/E_1)/(O_2/E_2)$, which will be close to e^X except in unusual data sets.

Worked example

For the data at the end of the study, shown in table 7.1, $O_1 = 10$, $E_1 = 11{\cdot}37$, $O_2 = 12$, $E_2 = 10{\cdot}63$, and $V = 4{\cdot}99$.

The values of X and Y for $\alpha = 0{\cdot}05$ are

$$X = \frac{10 - 11{\cdot}37}{4{\cdot}99} = -0{\cdot}28 \quad \text{and} \quad Y = \frac{1{\cdot}96}{\sqrt{4{\cdot}99}} = 0{\cdot}88.$$

The hazard ratio is thus estimated as

$$h = e^{-0{\cdot}28} = 0{\cdot}76.$$

The 95% confidence interval for the population value of the hazard ratio is then given by

$$e^{-1{\cdot}15} \quad \text{to} \quad e^{0{\cdot}60}$$

that is, from 0·32 to 1·83.

The results indicate that treatment with γ linolenic acid has been associated with an estimated reduction in mortality to 76% of that for the control treatment, while the more usual hazard ratio calculation gives a similar figure of 78%. This reduction, however, is imprecisely estimated as shown by the wide confidence interval of 32% to 183%.

Comment

Further discussion and examples are given by Simon.[6] He shows also how to calculate a confidence interval for the median survival time, which is a less commonly used statistic.

1 Pocock SJ. *Clinical trials: a practical approach*. Chichester: Wiley, 1983:222–4.
2 Peto R, Pike MC, Armitage P, *et al.* Design and analysis of randomized clinical trials requiring prolonged observation of each patient. II. Analysis and examples. *Br J Cancer* 1977;**35**:1–39.
3 Peto J. The calculation and interpretation of survival curves. In: Buyse ME, Staquet MJ,

Sylvester RJ, eds. *Cancer clinical trials: methods and practice*. Oxford: University Press, 1984:361–80.

4 McIllmurray MB, Turkie W. Controlled trial of γ linolenic acid in Dukes's C colorectal cancer. *Br Med J* 1987;**294**:1260 and **295**:475.

5 Daly L. Confidence intervals. *Br Med J* 1988;**297**:66.

6 Simon R. Confidence limits for reporting results of clinical trials. *Ann Intern Med* 1986;**105**:429–35.

8

Calculating confidence intervals for some non-parametric analyses

MICHAEL J CAMPBELL, MARTIN J GARDNER

The rationale behind the use of confidence intervals was described in chapters 1 and 2 and methods for their calculation for a population mean and for differences between two population means for paired and unpaired samples were given in chapter 3. These methods are based on sample means, standard errors, and the t distribution and should strictly be used only for continuous data from Normal distributions (although small deviations from Normality are not important[1]).

For non-Normal continuous data the median of the population or the sample is preferable to the mean as a measure of location. Medians are also appropriate in other situations—for example, when measurements are on an ordinal scale.

This chapter describes methods of calculating confidence intervals for a population median or for other population quantiles from a sample of observations. Calculations of confidence intervals for the difference between two population medians or means (a non-parametric approach rather than the parametric approach mentioned above) for both unpaired and paired samples are described. Worked examples are given for each situation.

Because of the discrete nature of some of the sampling distributions involved in non-parametric analyses it is not usually possible to calculate confidence intervals with exactly the desired level of confidence. Hence, if a 95% confidence interval is wanted the choice is between the lowest possible level of confidence over 95% (a "conservative" interval) and the highest possible under 95%. There is no firm policy on which of these is preferred, but we shall mainly describe conservative intervals in this chapter. The exact level of confidence associated with any particular approximate level can be calculated from the distribution of the statistic being used.

The methods outlined for obtaining confidence intervals are described in more detail in some textbooks on non-parametric statistics (for example, by Conover[2]). If there are many "ties" in the data—that is, observations with the same numerical value—then modifications to the formulae given here are needed.[2] The calculations can be carried out using some statistical computer packages such as MINITAB.[3] A method for calculating confidence intervals for Spearman's rank correlation coefficient is given in chapter 5.

A confidence interval indicates the precision of the sample statistic as an estimate of the overall population value. Confidence intervals convey the effects of sampling variation but cannot control for non-sampling errors in study design or conduct. They should not be used for basic description of the sample data but only for indicating the uncertainty in sample estimates for population values of medians or other statistics.

Confidence intervals for medians and other quantiles

Median

The median is defined as the value having half of the observations less than and half exceeding it. The sample median is used as an estimate of the population median. To find the $100(1-\alpha)\%$ confidence interval for the population median first calculate the quantities

$$r = \frac{n}{2} - \left(N_{1-\alpha/2} \times \frac{\sqrt{n}}{2}\right) \quad \text{and} \quad s = 1 + \frac{n}{2} + \left(N_{1-\alpha/2} \times \frac{\sqrt{n}}{2}\right),$$

where n is the sample size and $N_{1-\alpha/2}$ is the appropriate value from the standard Normal distribution for the $100(1-\alpha/2)$ percentile. This is widely available in tables—for example, table 1 in part III. Then round r and s to the nearest integers. The n sample observations need to be ranked in increasing order of magnitude and the rth to sth observations in the ranking then determine the $100(1-\alpha)\%$ confidence interval for the population median. This approximate method is satisfactory for most sample sizes.[4] The exact method, based on the Binomial distribution, can be used instead for small samples, as shown in the example below, which uses table 4 in part III. Alternatively, if the population distribution from which the observations came can be assumed to be symmetrical around the median the method described in the section on *Two samples: paired case* can be used.

Other quantiles

A similar approach can be used to calculate confidence intervals for quantiles other than the median—for example, the 90th percentile, which divides the lower nine tenths from the upper tenth of the observations. For the qth quantile ($q = 0{\cdot}9$ for the 90th percentile) r and s above are replaced with r' and s' given by

$$r' = nq - [N_{1-\alpha/2} \times \sqrt{nq(1-q)}]$$

and

$$s' = 1 + nq + [N_{1-\alpha/2} \times \sqrt{nq(1-q)}].$$

Worked examples

(1) For the data in the first example given in chapter 3 the median systolic blood pressure among 100 diabetic patients was 146 mm Hg. Using the above formulae to calculate a 95% confidence interval gives

$$r = \frac{100}{2} - \left(1{\cdot}96 \times \frac{\sqrt{100}}{2}\right) = 40{\cdot}2 \text{ and } s = 1 + \frac{100}{2} + \left(1{\cdot}96 \times \frac{\sqrt{100}}{2}\right) = 60{\cdot}8.$$

From the original data the 40th observation in increasing order is 142 mm Hg and the 61st is 150 mm Hg. The 95% confidence interval for the population median is thus from 142 to 150 mm Hg.

(2) For a small sample example using the Binomial distribution consider the results of a study measuring β endorphin concentrations in 11 subjects who had collapsed while running in a half marathon.[5] The concentrations in pmol/l in order of increasing value were: 66·0, 71·2, 83·0, 83·6, 101, 107·6, 122, 143, 160, 177, and 414.

The sample median is the 6th observation (107·6 pmol/l). To find a confidence interval for the population median we use the Binomial distribution[6] with probability of 0·5. For a conservative 95% confidence interval we first find the largest cumulative probability under 0·025 and the smallest over 0·975 either by direct calculation or from tables[7]. With $n = 11$ this gives Prob ($X \leq 1$) = 0·006 and Prob ($X \leq 9$) = 0·994.

The approximate 95% confidence interval is then found by the ranked observations that are one greater than those associated with the two probabilities—that is, the 2nd and 10th observation, giving 71·2 to 177 pmol/l. The actual probability associated with this confidence interval is 0·994 − 0·006 = 0·988, so effectively it is a 98·8% confidence interval. For sample sizes up to 100, the required rankings for approximate 90%, 95%, and 99% confidence intervals and associated exact levels of confidence are given directly in table 4 in part III.

Alternatively a non-conservative approximate 95% confidence interval can be found by calculating the smallest cumulative probability over 0·025 and the largest under 0·975. In this case $\text{Prob}(X \leqslant 2) = 0{\cdot}033$ and $\text{Prob}(X \leqslant 8) = 0{\cdot}967$, which give a 93·5% confidence interval from the 3rd to the 9th ranked observations—that is, from 83·0 to 160 pmol/l. In this case the coverage probability is nearer to 95% than for the conservative interval.

For this example the approximate large sample method gives the same result as the conservative 95% confidence interval.

Confidence intervals for differences between medians

In finding confidence intervals for population differences between medians it is assumed that the data come from distributions that are identical in shape and differ only in location. Because of this assumption the non-parametric confidence intervals described below can be regarded as being either for the difference between the two medians, or the difference between the two means, or the difference between any other two measures of location such as a particular percentile. This assumption is not necessary for a valid test of the null hypothesis of no difference in population distributions but if it is not satisfied the interpretation of a statistically significant result is difficult.

Two samples: unpaired case

Let $x_1, x_2, \ldots, x_{n_1}$ represent the n_1 observations in a sample from one population and $x'_1, x'_2, \ldots, x'_{n_2}$ the n_2 observations on the same variable in a sample from a second population, where both sets of data are thought not to come from Normal distributions. The difference between the two population medians or means is estimated by the median of all the possible $n_1 \times n_2$ differences $x_i - x'_j$ (for $i = 1$ to n_1 and $j = 1$ to n_2).

The confidence interval for the difference between the two population medians or means is also derived through these $n_1 \times n_2$ differences.[8] For an approximate $100(1 - \alpha)\%$ confidence interval first calculate

$$K = W_{\alpha/2} - \frac{n_1(n_1 + 1)}{2},$$

where $W_{\alpha/2}$ is the $100\alpha/2$ percentile of the distribution of the Mann–Whitney test statistic[9] or of the distribution of the equivalent Wilcoxon two sample test statistic[10] (in using the table in ref 10 take the lower of the two tabulated values and add 1 to it). The Kth smallest to the Kth largest of the $n_1 \times n_2$ differences then determine the $100(1-\alpha)\%$ confidence interval. Values of K for finding approximate 90%, 95%, and 99% confidence intervals ($\alpha = 0{\cdot}10$, $0{\cdot}05$, and $0{\cdot}01$), together with the associated exact levels of confidence, are given directly for sample sizes of up to 25 in table 5 in part III.

For studies where each sample size is greater than about 25, special tables are not required and K can be calculated approximately[2] as

$$K = \frac{n_1 n_2}{2} - \left(N_{1-\alpha/2} \times \sqrt{\frac{n_1 n_2 (n_1 + n_2 + 1)}{12}}\right),$$

rounded up to the next integer value, where $N_{1-\alpha/2}$ is the appropriate value from the standard Normal distribution for the $100(1-\alpha/2)$ percentile.

Worked example

Consider the data in table 8.1 on the globulin fraction of plasma (g/l) in two groups of 10 patients given by Swinscow.[11] The computations are

TABLE 8.1—Globulin fraction of plasma (g/l) in two groups of 10 patients[11]

Group 1	38	26	29	41	36	31	32	30	35	33
Group 2	45	28	27	38	40	42	39	39	34	45

TABLE 8.2—Differences in globulin fraction of plasma (g/l) between individuals in two groups of 10 patients[11]

	Group 1									
Group 2	26	29	30	31	32	33	35	36	38	41
27	−1	2	3	4	5	6	8	9	11	14
28	−2	1	2	3	4	5	7	8	10	13
34	−8	−5	−4	−3	−2	−1	1	2	4	7
38	−12	−9	−8	−7	−6	−5	−3	−2	0	3
39	−13	−10	−9	−8	−7	−6	−4	−3	−1	2
39	−13	−10	−9	−8	−7	−6	−4	−3	−1	2
40	−14	−11	−10	−9	−8	−7	−5	−4	−2	1
42	−16	−13	−12	−11	−10	−9	−7	−6	−4	−1
45	−19	−16	−15	−14	−13	−12	−10	−9	−7	−4
45	−19	−16	−15	−14	−13	−12	−10	−9	−7	−4

made easier if the data in each group are first ranked into increasing order of magnitude and then all the group 1 minus group 2 differences calculated as in table 8.2. The estimate of the difference in population medians or means is now given by the median of these differences. From the 100 differences in table 8.2 the 50th smallest difference is -6 g/l and the 51st is -5 g/l, so the median difference is estimated as $[-6+(-5)]/2 = -5{\cdot}5$ g/l.

To calculate an approximate 95% confidence interval for the difference in population medians or means the value of $K=24$ is found for $n_1=10$, $n_2=10$, and $\alpha=0{\cdot}05$ from table 5 in part III. The 24th smallest difference is -10 g/l and the 24th largest is $+1$ g/l. The approximate 95% confidence interval (exact level 95·7%) for the difference in population medians or means is thus from -10 g/l to $+1$ g/l.

Two samples: paired case

Paired cases include studies of repeated measurements on the same individuals and matched case-control comparisons where paired differences are the observations of main interest. The method for finding confidence intervals described here assumes that, as well as the two distributions being identical except in location, the distribution of the paired differences is symmetrical. If this additional assumption seems unreasonable then the method described previously for a single sample can be applied to the paired differences.

Suppose that in a sample of size n the differences for each matched pair of measurements are $d_1, d_2, \ldots, d_n$. The difference between the two population medians or means is estimated by calculating all the $n(n+1)/2$ possible averages of two of these differences taken together, including each difference with itself, and selecting their median.

The confidence interval for the difference between the population medians or means is also derived using these averaged differences.[12] For an approximate $100(1-\alpha)$% confidence interval first find the value of $W_{\alpha/2}$ as the $100\alpha/2$ percentile of the distribution of the Wilcoxon one sample test statistic[13 14] (in using the table in ref 14 take the lower of the two tabulated values and add 1 to it). Then if $W_{\alpha/2}=K^\star$ the $K^\star$th smallest to the $K^\star$th largest of the averaged differences determine the $100(1-\alpha)$% confidence interval. Values of $K^\star$ for finding approximate 90%, 95%, and 99% confidence intervals ($\alpha=0{\cdot}10$, $0{\cdot}05$, and $0{\cdot}01$), together with the associated exact levels of confidence, are given directly for sample sizes of up to 50 in table 6 in part III. In general the degree of conservatism is small.

For sample sizes of about 50 or more special tables are not required and $K^\star$ can be calculated approximately[2] as

$$K^\star = \frac{n(n+1)}{4} - \left(N_{1-\alpha/2} \times \sqrt{\frac{n(n+1)(2n+1)}{24}}\right),$$

rounded up to the next integer value, where $N_{1-\alpha/2}$ is the appropriate value from the standard Normal distribution for the $100(1-\alpha/2)$ percentile.

Worked example

Consider further results of measuring β endorphin concentrations in subjects running in a half marathon where 11 people were studied both before and after the event.[5] The before and after concentrations (pmol/l) and their differences ordered by increasing size were as given in table 8.3.

TABLE 8.3—β endorphin concentrations in 11 runners before and after a half marathon[5]

Subject No	β endorphin concentration		Change in concentration
	Before	After	After–before
1	10·6	14·6	4·0
2	5·2	15·6	10·4
3	8·4	20·2	11·8
4	9·0	20·9	11·9
5	6·6	24·0	17·4
6	4·6	25·0	20·4
7	14·1	35·2	21·1
8	5·2	30·2	25·0
9	4·4	30·0	25·6
10	17·4	46·2	28·8
11	7·2	37·0	29·8

All the possible $n(n+1)/2$ averages in this case where $n = 11$ give the 66 averages shown in table 8.4. Thus, having found $K^\star = 11$ for $n = 11$ and $\alpha = 0{\cdot}05$ from table 6 in part III, the 11 smallest averages are 4·0, 7·2, 7·9, 7·95, 10·4, 10·7, 11·1, 11·15, 11·8, 11·85, and 11·9; and the 11 largest averages are 25·1, 25·3, 25·45, 25·6, 26·9, 27·2, 27·4, 27·7, 28·8, 29·3, and 29·8. The approximate 95% (exact 95·8%) confidence interval for the difference between the population medians or means is thus given as 11·9 to 25·1 pmol/l around the sample median which is 18·8 pmol/l (the

TABLE 8.4—Averages of differences in β endorphin concentrations in 11 runners before and after a half marathon[5]

	Change										
Change	4·0	10·4	11·8	11·9	17·4	20·4	21·1	25·0	25·6	28·8	29·8
4·0	4·00	7·20	7·90	7·95	10·70	12·20	12·55	14·50	14·80	16·40	16·90
10·4		10·40	11·10	11·15	13·90	15·40	15·75	17·70	18·00	19·60	20·10
11·8			11·80	11·85	14·60	16·10	16·45	18·40	18·70	20·30	20·80
11·9				11·90	14·65	16·15	16·50	18·45	18·75	20·35	20·85
17·4					17·40	18·90	19·25	21·20	21·50	23·10	23·60
20·4						20·40	20·75	22·70	23·00	24·60	25·10
21·1							21·10	23·05	23·35	24·95	25·45
25·0								25·00	25·30	26·90	27·40
25·6									25·60	27·20	27·70
28·8										28·80	29·30
29·8											29·80

average of the 33rd and 34th ranked observations, 18·75 and 18·90, in the table of average differences). The triangular table of average differences helps to identify the required values, but a computer package such as MINITAB[3] can rank the averages in order and select the appropriate ranked values.

Technical note

It should be noted that there are differences of presentation in the tables referred to in Conover,[2] *Geigy Scientific Tables*,[15] and elsewhere. These result from the discrete nature of the distributions and whether[15] or not[2] the tabulated values are part of the critical region for the test of the null hypothesis. MINITAB[3] uses large sample formulae with continuity corrections for computing the coverage probabilities of the confidence intervals for differences between medians even for sample sizes less than 20. This can lead to inaccuracies in the coverage probabilities given by the program.

1 Bland M. *An introduction to medical statistics*. Oxford: University Press, 1987:179–82.
2 Conover WJ. *Practical non-parametric statistics*. New York: Wiley, 1980.
3 Ryan BF, Joiner BL, Ryan TA. *Minitab handbook*. 2nd ed. Boston: Duxbury Press, 1985.
4 Hill ID. 95% confidence limits for the median. *J Statist Comput and Simulation* 1987;**28**:80–1.
5 Dale G, Fleetwood JA, Weddell A, Ellis RD, Sainsbury JRC. β Endorphin: a factor in "fun run" collapse. *Br Med J* 1987;**294**:1004.
6 Bland M. *An introduction to medical statistics*. Oxford: University Press, 1987: 98–101.
7 Conover WJ. *Practical non-parametric statistics*. New York: Wiley, 1980: table A3.
8 Conover WJ. *Practical non-parametric statistics*. New York: Wiley, 1980:223–5.
9 Conover WJ. *Practical non-parametric statistics*. New York: Wiley, 1980: table A7.

10 Lentner C, ed. *Geigy scientific tables*. Vol 2. 8th ed. Basle: Ciba-Geigy, 1982:156–62.
11 Swinscow TDV. *Statistics at square one*. 7th ed. London: British Medical Association, 1980:58–9.
12 Conover WJ. *Practical non-parametric statistics*. New York: Wiley, 1980:288–90.
13 Conover WJ. *Practical non-parametric statistics*. New York: Wiley, 1980: table A13.
14 Lentner C, ed. *Geigy scientific tables*. Vol 2. 8th ed. Basle: Ciba-Geigy, 1982:163.
15 Lentner C, ed. *Geigy scientific tables*. Vol 2. 8th ed. Basle: Ciba-Geigy, 1982.

Part II
Statistical guidelines and check lists

9

Statistical guidelines for contributors to medical journals

DOUGLAS G ALTMAN, SHEILA M GORE,
MARTIN J GARDNER, STUART J POCOCK

Introduction

Most papers published in medical journals contain analyses that have been carried out without any help from a statistician. Although nearly all medical researchers have some acquaintance with basic statistics, there is no easy way for them to acquire insight into important statistical concepts and principles. There is also little help available about how to design, analyse, and write up a whole project. Partly for these reasons much that is published in medical journals is statistically poor or even wrong.[1] A high level of statistical errors has been noted in several reviews of journal articles and has caused much concern.

Few journals offer even rudimentary statistical advice to contributors. It has been suggested[1 2] that comprehensive statistical guidelines could help by making medical researchers more aware of important statistical principles, and by indicating what information ought to be supplied in a paper. We present below an attempt to do this. Since our original article was written Bailar and Mosteller have published a set of guidelines amplifying the brief section on statistics in the "Uniform requirements for manuscripts."[3 4] Other authors have published guidelines for particular types of study.[5–7]

Deciding what to include in the guidelines, how much detail to give, and how to deal with topics where there is no consensus has been problematic. These guidelines should thus be seen as one view of what is important, rather than as a definitive document. We have not set out to provide a set of rules but rather to give general information and advice about important aspects of statistical design, analysis, and presentation. Those specific recommendations that we have made are mostly strong advice against certain practices.

Some familiarity with statistical methods and ideas is assumed, since some knowledge of statistics is necessary before carrying out statistical analyses. For those with only a limited acquaintance with statistics, the guidelines should show that the subject is very much wider than mere significance testing and illustrate how important correct interpretation is. The lack of precise recommendations in some places indicates that good statistical analysis requires common sense and judgment, as well as a repertoire of formal techniques, so that there is an art in statistics as well as in medicine. We hope that the guidelines present an uncontroversial view of the most frequently used and accepted statistical procedures. We have deliberately limited the scope of the guidelines to cover the more common statistical procedures.

Readers may find that a relevant section presents information or advice that is unfamiliar or is not understood. In such circumstances, although almost all of the topics covered may be found in the more comprehensive medical statistics textbooks,[8–11] we strongly recommend that they should seek the advice of a statistician. The absence from the guidelines of specific references is intentional: it is better to get expert personal advice if further insight is needed. Moreover, because mistakes in design cannot later be rectified, professional advice should first be obtained when planning a research project rather than when analysing the data.

These guidelines are intended to try to help authors know what is important statistically and how to present it in their papers. They emphasise that such matters of presentation are closely linked to more general consideration of statistical principles. Detailed discussion of how to choose an appropriate statistical method is not given; such information is best obtained by consulting a statistician. We do, however, draw attention to certain misuses of statistical methods.

These guidelines follow the usual structure of medical research papers: Methods, Results (analysis and presentation), and Discussion (interpretation). As a result several topics appear in more than one place and are cross referenced as appropriate. Statistical check lists (chapter 10) indicate the broad categories of information that should be included in a paper.

Methods section

General principles

It is most important to describe clearly what was done, including the design of the research (be it an experiment, trial, or survey) and the collection of the data. The aim should be to give enough information to allow methods to be fully understood and, if desired, repeated by others. Authors should include information on the following aspects of the design of their research:

- the objective of the research, and major hypotheses;
- the type of subjects, stating criteria for inclusion and exclusion;
- the source of the subjects and how they were selected;
- the number of subjects studied and why that number of subjects was used;
- the types of observation and the measurement techniques used (where several assessments are made for each subject, the main focus of interest should be specified).

Each type of study—for example, surveys and clinical trials—will require certain additional information.

Surveys (observational studies)

The study design should be clearly explained. For instance, the selection of a control group and any matching procedures need detailed description. It should also be clearly stated whether the study is retrospective, cross sectional, or prospective. The procedure for selecting subjects and the achievement of a high participation rate are particularly important, as findings are usually extrapolated from the sample to some general population. It is helpful to report any steps taken to encourage participation in the survey.

Clinical trials

The treatment regimens (including ancillary patient care and criteria for modifying or stopping treatment) need detailed definition. The method for allocating treatments to subjects should be stated explicitly. In particular, the specific method of randomisation (including any stratification) and how it was implemented need to be explained. Any lack of randomisation should be noted as a deficiency in design and the reasons given.

The use of blinding techniques and other precautions taken to ensure an unbiased evaluation of patient response should be described. The main criteria for comparing treatments, as agreed in the trial protocol, should be listed. For crossover trials the precise pattern of treatments (and any run in and wash out periods) needs explaining.

A more comprehensive list of information to include in the report of a clinical trial is given in the check list in chapter 10.

Statistical methods

All the statistical methods used in a paper should be identified. When several techniques are used it should be absolutely clear which method was used where, and this may need clarification in the results section. Common techniques, such as *t* tests, simple χ^2 tests, Wilcoxon and Mann–Whitney tests, correlation (*r*), and linear regression, do not need to be described, but methods with more than one form, such as *t* tests (paired or unpaired), analysis of variance, and rank correlation, should be identified unambiguously. More complex methods do need some explanation, and if the methods are unusual a precise reference should be given, preferably to a textbook (with page numbers). It may help to include brief comments on why the particular method of analysis was used, especially when a more familiar approach has been avoided. It is useful to give the name of a computer program or package used—for example, the Statistical Package for the Social Sciences (SPSS)—but the specific statistical methods must still be identified.

Results section: statistical analysis

Descriptive information

Adequate description of the data should precede and complement formal statistical analysis. In general variables which are important for the validity and interpretation of subsequent statistical analyses should be described in most detail. This can be achieved by graphical methods, such as scatter plots or histograms, or by using summary statistics. Continuous variables (such as weight or blood pressure) can be summarised using the mean and standard deviation (SD) or the median and a percentile range—say, the interquartile range (25th to 75th percentile). The latter approach is preferable when continuous measurements have an asymmetrical distribution. For ordered qualitative data (such as stages of disease I to IV) the calculation of means and standard deviations is incorrect; instead, proportions should be reported.

Deviations from the intended study design should be described. For example, in clinical trials it is particularly important to enumerate withdrawals with reasons, if known, and treatment allocation. For surveys, where the response rate is of fundamental importance, it is valuable to give information on the characteristics of the non-responders compared with those who took part. The representativeness of the study sample will need to be investigated if it is a prime intention to extrapolate results to some appropriate population.

It is useful to compare the distribution of bascline characteristics in different groups, such as treatment groups in a clinical trial. Such differences that exist, even if not statistically significant, are real and should be properly allowed for in the analysis (see *Complex analyses*).

Underlying assumptions

Methods of analysis such as t tests, correlation, regression, and analysis of variance all depend to some extent on certain assumptions about the distribution of the variable(s) being analysed. Technically, these assumptions are that in some aspect the data come from a Normal distribution and if two or more groups are being compared that the variability within each is the same.

It is not possible to give absolutely the degree to which these assumptions may be violated without invalidating the analysis. But data which have a highly skewed (asymmetrical) distribution or for which the variability is considerably different across groups may require either some transformation before analysis (see *Data transformation*) or the use of alternative "distribution free" methods, which do not depend on assumptions about the distribution (often called non-parametric methods). For example, the Mann–Whitney U test is the distribution free equivalent of the two sample t test. Distribution free methods may also be appropriate for small data sets, for which the assumptions cannot be validated adequately.

Sometimes the assumption of Normality may be especially important—for example, when the range of values calculated as two standard deviations either side of the mean is taken as a 95% "normal" or reference range. In such cases the distributional assumption must be shown to be justified.

Hypothesis tests

The main purpose of hypothesis tests (often less accurately referred to as significance tests) is to evaluate a limited number of preformulated hypotheses. Other tests, which are carried out because they have been suggested by preliminary inspection of the data, will give a false impression because in such circumstances the calculated P value is too small. For example, it is not valid to test the difference between the smallest and largest of a set of several means or proportions without making due allowance for the reason for testing that particular difference; special "multiple comparison" techniques are available for making pairwise comparisons among several groups. However, where three or more groups are compared which have a natural ordering, such as age groups or stages of cancer, the data should be analysed by a method that specifically evaluates a trend across groups.

It is customary to carry out two sided hypothesis tests. If a one sided test is used this should be indicated and justified for the problem in hand.

The presentation and interpretation of results of hypothesis tests are discussed in later sections. The use of confidence intervals in addition to hypothesis tests is strongly recommended—see next section and chapters 1 and 2.

Confidence intervals

Most studies are concerned with estimating some quantity, such as a mean difference or a relative risk. It is desirable to calculate the confidence interval around such an estimate. This is a range of values about which we are, say, 95% confident that it includes the true value. There is a close relation between the results of a test of a hypothesis and the associated confidence interval: if the difference between treatments is significant at the 5% level then the associated 95% confidence interval excludes the zero difference. The confidence interval conveys more information because it indicates a range of values for the true effect which is compatible with the sample observations (see also *Interpretation of hypothesis tests* and chapter 2).

Confidence intervals reveal the precision of an estimate. A wide confidence interval points to lack of information, whether the difference is statistically significant or not, and is a warning against overinterpreting results from small studies.

In a comparative study confidence intervals should be reported for the differences between groups, not for the results of each group separately.

Paired observations

It is essential to distinguish the case of unpaired observations, where the comparison is between measurements for two different groups—for example, subjects receiving alternative treatments—from that of paired observations, where the comparison is between two measurements made on the same individuals in different circumstances (such as before and after treatment). For example, where with unpaired data the two sample *t* test would be used, with paired data the paired *t* test should be used instead. Similarly, the Mann–Whitney U test for unpaired data is replaced by the paired Wilcoxon test, and the usual χ^2 test for 2×2 tables is replaced by McNemar's test. It should always be made clear which form of test was used. Likewise the method for calculating a confidence interval differs from that for unpaired observations (see chapters 3, 4, 6, and 8).

The same distinction must be made when there are three or more sets of observations. All of the statistical methods mentioned in this section may be generalised to more than two groups; in particular, paired and two sample *t* tests generalise to different forms of analysis of variance.

Repeated measurements

A common study design entails recording serial measurements of the same variable(s) on the same individual at several points in time. Such data are often analysed by calculating means and standard deviations at each time and presented graphically by a line joining these means. The shape of this mean curve may not give a good idea of the shapes of the individual curves. Unless the individual responses are very similar it may be more valuable to analyse some characteristic of the individual profiles, such as the time taken to reach a peak or the length of time above a given level. This would also help to avoid the problems associated with multiple hypothesis testing (see *Many hypothesis tests*).

Repeated measurements of the same variable on one individual under the same experimental conditions, known as replicate readings, should not be treated as independent observations when comparing groups of individuals. Where the number of replicates is the same for all subjects analysis is not difficult; in particular, analysis of variance is used where *t* tests would have been applied to unreplicated data. If the number of replicates varies among individuals a full analysis can be very complex. The use of the largest or smallest of a series of measurements (such as maximum blood pressure during pregnancy) may be misleading if the number of observations varies widely among individuals.

Data transformation

Many biomedical variables have distributions which are positively skewed, with some very high values, and they may require mathematical transformation to make the data appropriate for analysis. In such circumstances the logarithmic (log) transformation is often applicable, although occasionally other transformations (such as square root or reciprocal) may be more suitable.

After analysis it is desirable to convert the results back into the original scale for reporting. In the common case of log transformation the antilog of the mean of the log data (known as the geometric mean) should be used. The standard deviation or standard error must not be antilogged, however; instead, a confidence interval on the log scale can be antilogged to give a confidence interval on the original scale. A similar procedure is adopted with other transformations when there is a single sample, but back transformation of the confidence interval for a difference between sample means makes sense only for the log transformation (see chapter 3).

If a transformation is used it is important to check that the desired effect (such as an approximately Normal distribution) is achieved. It should not be assumed that the log transformation, for instance, is necessarily suitable for all positively skewed variables.

Outliers

Observations that are highly inconsistent with the main body of the data should not be excluded from the analysis unless there are additional reasons to doubt their credibility. Any omission of such outliers should be reported. Because outliers can have a pronounced effect on a statistical analysis it is useful to analyse the data both with and without such observations to assess how much any conclusions depend on these values.

Correlation

It is preferable to include a scatter plot of the data for each correlation coefficient presented, although this may not be possible if there are several variables. When many variables are being investigated it is useful to show the correlations between all pairs of variables in a table (correlation matrix), rather than quoting just the largest or significant values.

For data which are irregularly distributed the rank correlation can be calculated instead of the usual Pearson "product moment" correlation (r). Rank correlation can also be used for variables that are constrained to be above or below certain values—for example, birth weights below 2500 g—or for ordered categorical variables. Rank correlation is also preferable when the relation between the variables is not linear, or when the values of one variable have been chosen by the experimenter rather than being unconstrained.

The correlation coefficient is a useful summary of the degree of linear association between two quantitative variables, but it is one of the most misused statistical methods. There are several circumstances in which correlation ought not to be used. It is incorrect to calculate a simple correlation coefficient for data which include more than one observation on some or all of the subjects, because such observations are not independent. Correlation is inappropriate for comparing alternative methods of measurement of the same variable because it assesses association, not agreement. The use of correlation to relate change over time to the initial value can give grossly misleading results.

It may be misleading to calculate the correlation coefficient for data comprising subgroups known to differ in their mean levels of one or both variables—for example, combining data for men and women when one of the variables is height.

Regression and correlation are separate techniques serving different purposes and need not automatically accompany each other. The interpretation of correlation coefficients is discussed below (*Association and causality*).

Regression

It is highly desirable to present a fitted regression line together with a scatter diagram of the raw data. A plot of the fitted line without the data gives little futher information than the regression equation itself. It is useful to give the values of the slope (with its standard error) and intercept and a measure of the scatter of the points around the fitted line (the residual standard deviation). A confidence interval may be constructed for a regression line and prediction intervals for individuals based on the fitted relationship. The lines joining these values are not parallel to the regression line but curved, showing the greater uncertainty of the prediction corresponding to values on the horizontal (x) axis away from the bulk of the observations (see chapter 5).

Regression on data including distinct subgroups can give misleading results, particularly if the groups differ in their mean level of the dependent (y) variable. More reliable results may be obtained by using analysis of covariance.

Regression and correlation are separate techniques serving different purposes and need not automatically accompany each other. The interpretation of regression analysis is discussed below (*Prediction and diagnostic tests*).

Survival data

The reporting of survival data should include graphical or tabular presentation of life tables, with details of how many patients were at risk (of dying, say) at different follow up times (see chapter 7). The life table or actuarial survival curve deals efficiently with the "censored" survival times which arise when patients are lost to follow up or are still alive; their survival time is known to be only at least so many days. To avoid misinterpretation of the unreliable later part of the curve it may help to truncate the survival curve when there are only a few (say 5) subjects still at risk. The calculation of mean survival time is inadvisable in the presence of censoring and because the distribution of survival times is usually positively skewed.

Comparison between treatment groups of the proportion surviving at arbitrary fixed times can be misleading and is generally less efficient than the comparison of life tables by a method such as the logrank test. Methods for calculating estimates of survival and confidence intervals are given in chapter 7.

When there are sufficient deaths one can show how the risk of dying varies with time by plotting, for suitable equal time intervals, the proportion of those alive at the beginning of each time interval who died during that interval. Adjusting for patient factors which might influence prognosis is possible using regression models appropriate to survival data (see next section).

Comparison of survival between the group of individuals who respond to treatment and the group who do not is misleading and should never be performed.

Complex analyses

In many studies the observations of prime interest may be influenced by several other variables. These might be anything that varies among subjects and which might have affected the outcome being observed. For example, in clinical trials they might include patient characteristics or signs and symptoms. Some or all of the covariates can be combined by appropriate multiple regression techniques to explain or predict an outcome variable, be it a continuous variable (blood pressure), a qualitative variable (post-operative thrombosis), or the length of survival. Even in randomised clinical trials investigators need assurance that the treatment effect is still present after simultaneous adjustment for several risk factors.

Multivariate techniques, for dealing with more than one outcome variable simultaneously, really require expert help and are beyond the scope of these guidelines.

Any complex statistical methods should be communicated in a manner that is comprehensible to the reader.

Results section: presentation of results

Presentation of summary statistics

Mean values should not be quoted without some measure of variability or precision. The standard deviation (SD) should be used to show the variability among individuals and the standard error of the mean (SE) to show the precision of the sample mean (see chapter 2: appendix 1). It must be made clear which is presented.

The use of the symbol ± to attach the standard error or standard deviation to the mean (as in 14·2 ± 1·9) causes confusion and should be avoided. Several medical journals do not now allow its use. The presentation of means as, for example, 14·2 (SE 1·9) or 14·2 (SD 7·4) is preferable. Confidence intervals are a good way of providing a reasonable indication of uncertainty of sample means, proportions, and other statistics. For example, a 95% confidence interval for the true mean is from about two standard errors below the observed mean to two standard errors above it (see chapter 3). Confidence intervals are more clearly presented as 10·4 to 18·0 (see chapter 2) than by use of the ± symbol.

When paired comparisons are made, such as when using paired *t* tests, it is important to give the mean and standard deviation of the differences between the observations or the standard error of the mean difference as appropriate (see chapter 2: appendix 1).

For data that have been analysed with distribution free methods it is more appropriate to give the median and a central range, covering, for example, 95% of the observations, than to use the mean and standard deviation (see *Descriptive information*). Non-parametric confidence intervals can be calculated (see chapter 8). Likewise, if analysis has been carried out on transformed data the mean and standard deviation of the raw data will probably not be good measures of the centre and spread of the data and should not be presented.

When percentages are given the denominator should always be made clear. For small samples the use of percentages is unhelpful. When percentages are contrasted it is important to distinguish an absolute difference from a relative difference. For example, a reduction from 25% to 20% may be expressed as either 5% or 20%.

Results for individuals

The overall range is not a good indicator of the variability of a set of observations as it can be strongly affected by a single extreme value and it increases with sample size. If the data have a reasonably Normal distribution the interval two standard deviations either side of the mean will cover about 95% of the observations, but a percentile range is more widely applicable to other distributions (see *Descriptive information*).

Although statistical analysis is concerned with average effects, in many circumstances it is important also to consider how individual subjects responded. Thus, for example, it is very often clinically relevant to know how many patients did not improve with a treatment as well as the average benefit. An average effect should not be interpreted as applying to all individuals (see also *Repeated measurements*).

Presentation of results of hypothesis tests

Hypothesis tests yield observed values of test statistics which are compared with tabulated values for the appropriate distribution

(Normal, t, χ^2, etc) to derive associated P values. It is desirable to report the observed values of the test statistics and not just the P values. The quantitative results being tested, such as mean values, proportions, correlation coefficients, should be given whether the test was significant or not. It should be made clear precisely which data have been analysed. If symbols, such as asterisks, are used to denote levels of probability these must be defined and it is helpful if they are the same throughout the paper.

P values are conventionally given as <0·05, <0·01, or <0·001, but there is no reason other than familiarity for using these particular values. Exact P values (to no more than two significant figures), such as P = 0·18 or 0·03, are more helpful. It is unlikely to be necessary to specify levels of P lower than 0·0001. Calling any value with P > 0·05 "not significant" is not recommended, as it may obscure results that are not quite statistically significant but do suggest a real effect (see *Interpretation of hypothesis tests*). When quoting P values it is important to distinguish < (less than) from > (greater than). P values between two limits should be expressed in logical order—for example, 0·01 < P < 0·05 where P lies between 0·01 and 0·05. P values given in tables need not be repeated in the text.

The interpretation of hypothesis tests and P values is discussed below (*Interpretation of hypothesis tests*).

Figures (graphical presentation)

Graphical display of results is helpful to readers, and figures that show individual observations are to be encouraged. Points on a graph relating to the same individual on different occasions should preferably be joined, or symbols used to indicate the related points. A helpful alternative is to plot the difference between occasions for each individual.

The customary "error bars" of one standard error above and below the mean depict only a 67% confidence interval, and are thus liable to misinterpretation; 95% confidence intervals are preferable. The presentation of such information in figures is subject to the same considerations as discussed above (*Presentation of summary statistics*).

Scatter diagrams relating two variables should show all the observations, even if this means slight adjustment to accommodate duplicate points. These may also be indicated by replacing the plotting symbol by the actual number of coincident points.

Tables

It is much easier to scan numerical results down columns rather than across rows, and so it is better to have different types of information (such as means and standard errors) in separate columns. The number of observations should be stated for each result in a table. Tables giving information about individual patients, geographical areas, and so on are easier to read if the rows are ordered according to the level of one of the variables presented.

Numerical precision

Spurious precision adds no value to a paper and even detracts from its readability and credibility. Results obtained from a calculator or computer usually need to be rounded. When presenting means, standard deviations, and other statistics the author should bear in mind the precision of the original data. Means should not normally be given to more than one decimal place more than the raw data, but standard deviations or standard errors may need to be quoted to one extra decimal place. It is rarely necessary to quote percentages to more than one decimal place, and even one decimal place is often not needed. With samples of less than 100 the use of decimal places implies unwarranted precision and should be avoided. Note that these remarks apply only to presentation of results—rounding should not be used before or during analysis. It is sufficient to quote values of t, χ^2, and r to two decimal places.

Miscellaneous technical terms

It is impossible to define here all statistical terms. The following comments relate to some terms which are frequently used in an incorrect or confusing manner.

Correlation should preferably not be used as a general term to describe any relationship. It has a specific technical meaning as a measure of association, for which it should be reserved in statistical work.

Incidence should be used to describe the rate of occurrence of new cases of a given characteristic in a study sample or population, such as the number of new notifications of cancer in one year. The proportion of a sample already having a characteristic is the *prevalence*.

Non-parametric refers to certain statistical analyses, such as the

Mann–Whitney U test; it is not a characteristic of the observations themselves.

Parameter should not be used in place of "variable" to refer to a measurement or attribute on which observations are made. Parameters are characteristics of distributions or relationships in the population which are estimated by statistical analysis of a sample of observations.

Percentiles—When the range of values of a variable is divided into equal groups, the cut off points are the median, tertiles, quartiles, quintiles, and so on; the groups themselves should be referred to as halves, thirds, quarters, fifths, etc.

Sensitivity is the ability of a test to identify a disease when it really is present—that is, the proportion positive of those who have the disease. *Specificity* is the ability of a test to identify the absence of a disease when the disease really is not present—that is, the proportion negative of those who do not have the disease. See also *Prediction and diagnostic tests*.

Discussion section: interpretation

Interpretation of hypothesis tests

A hypothesis test assesses, by means of the probability P, the plausibility of the observed data when some "null hypothesis" (such as there being no difference between groups) is true. The P value is the probability that the observed data, or a more extreme outcome, would have occurred by chance—that is, just due to sampling variation—when the null hypothesis is true. If P is small one doubts the null hypothesis. If P is large the data are plausibly consistent with the null hypothesis, which thus cannot be rejected. P is not, therefore, the probability of there being no real effect.

Even if there is a large real effect a non-significant result is quite likely if the number of observations is small. Conversely, if the sample size is very large a statistically significant result may occur when there is only a small real effect. Thus statistical significance should not be taken as synonymous with clinical importance.

The interpretation of the results of hypothesis tests largely follows from the above. A significant result does not necessarily indicate a real effect. There is always some risk of a false positive finding; this risk diminishes for smaller P values. Furthermore, a non-significant result (conventionally $P>0{\cdot}05$) does not mean that there is no effect but only that the data are compatible with there being no effect. Some flexibility is desirable in interpreting P values. The 0·05 level is a convenient cut off point, but P values of 0·04 and 0·06, which are not greatly different, ought to lead to similar interpretations, rather than radically different ones. The designation of any result with P > 0·05 as not significant may thus mislead the reader (and the authors); hence the suggestion above (*Presentation of results of hypothesis tests*) to quote actual P values.

Confidence intervals are extremely helpful in interpretation, particularly for small studies, as they show the degree of uncertainty related to a result—such as the difference between two means—whether or not it was statistically significant. Their use in conjunction with non-significant results may be especially enlightening.

Many hypothesis tests

In many research projects some tests of hypotheses relate to important comparisons that were envisaged when the research was initiated. Tests of hypotheses which were not decided in advance are subsidiary, especially if suggested by the results. It is important to distinguish these two cases and give much greater weight to the tests of those hypotheses that were formulated initially. Other tests should be considered as being only exploratory—for forming new hypotheses to be investigated in further studies. One reason for this is that when very many hypothesis tests are performed in the analysis of one study, perhaps comparing many subgroups or looking at many variables, a number of spurious positive results can be expected to arise by chance alone, which may pose considerable problems of interpretation. Clearly, the more tests that are carried out the greater is the likelihood of finding some significant results, but the expected number of false positive findings will increase too. One way of allowing for the risk of false positive results is to set a smaller level of P as a criterion of statistical significance.

A more complex problem arises when tests of significance are carried out on dependent (correlated) data. One example of this is in the analysis of serial data (discussed above—*Repeated measurements*), when the same test is performed on data for the same variable collected from the same subjects at different times. Another is where separate analyses of two or more correlated variables are carried out as if they were independent; any corroboration may not greatly increase the weight of evidence because the tests relate to similar data. For example, diastolic and systolic blood pressures behave very similarly, as may alternative ways of assessing patient response generally. Very careful interpretation of results is required in such cases.

Association and causality

A statistically significant association (obtained from correlation or χ^2 analysis) does not in itself provide direct evidence of a causal relationship between the variables concerned. In observational studies causality can be established only on non-statistical grounds; it is easier to infer causality in randomised trials. Great care should be taken in comparing variables which both vary with time, because it is easy to obtain apparent associations which are spurious.

Prediction and diagnostic tests

Even when regression analysis has indicated a statistically significant relationship between two variables there may be considerable imprecision when using the regression equation to predict the numerical level of one variable (y) from the other (x) for individual cases. The accuracy of such predictions cannot be assessed from the correlation or regression coefficient but requires the calculation of the prediction interval for the estimated y value corresponding to a specific x value (see chapter 5). The regression line should be used only to predict the y variable from the x variable, and not the reverse.

A diagnostic test with a high sensitivity and specificity may not necessarily be a useful test for diagnostic purposes, especially when applied in a population where the prevalence of the disease is very low. It is useful here to calculate the proportion of subjects with positive test results who actually had the disease (known as the positive predictive value). Note that there is no consensus on the definition of "false positive rate" or "false negative rate"; it should always be made clear exactly what is being calculated, and this can best be illustrated by a 2×2 table relating the test results to the patients' true disease status.

A similar diagnostic problem arises with continuous variables. The classification as "abnormal" of values outside the "normal range" for a variable is common, but if the prevalence of true abnormality is low most values outside the normal range will be normal. The definition of abnormality should be based on both clinical and statistical criteria.

Weaknesses

It is better to address weaknesses in research design and execution, if one is aware of them, and to consider their possible effects on the results and their interpretation than to ignore them in the hope that they will not be noticed.

Concluding remarks

The purpose of statistical methods is to provide a straightforward factual account of the scientific evidence derived from a piece of research. The skills and experience needed to design suitable studies, carry out sensible statistical analyses, and communicate the findings in a clear and objective manner are not easy to acquire. We hope that these guidelines may contribute to an improvement in the standard of statistical work reported in medical publications.

1 Altman DG. Statistics in medical journals. *Statistics in Medicine* 1982;**1**:59–71.
2 O'Fallon JR, Dubey SB, Salsburg DS, *et al.* Should there be statistical guidelines for medical research papers? *Biometrics* 1978;**34**:687–95.
3 Bailar JC, Mosteller F. Guidelines for statistical reporting in articles for medical journals: amplifications and explanations. *Ann Intern Med* 1988;**108**:266–73.
4 International Committee of Medical Journal Editors. Uniform requirements for manuscripts submitted to biomedical journals. *Br Med J* 1988;**296**:401–5.
5 Simon R, Wittes RE. Methodologic guidelines for reports of clinical trials. *Cancer Treat Rep* 1985;**69**:1–3.
6 Epidemiology Work Group of the Interagency Regulatory Liaison Group. Guidelines for the documentation of epidemiologic studies. *Am J Epidemiol* 1981;**114**:609–13.
7 Lichtenstein MJ, Mulrow CD, Elwood PC. Guidelines for reading case-control studies. *J. Chron Dis* 1987;**40**:893–903.
8 Armitage P, Berry G. *Statistical methods in medical research.* 2nd ed. Oxford: Blackwell, 1987.
9 Colton T. *Statistics in medicine.* Boston: Little, Brown, 1974.
10 Bland M. *An introduction to medical statistics.* Oxford: University Press, 1987.
11 Bradford Hill A. *A short textbook of medical statistics.* 11th ed. London: Hodder and Stoughton, 1984.

10

Use of check lists in assessing the statistical content of medical studies

MARTIN J GARDNER, DAVID MACHIN
MICHAEL J CAMPBELL

Summary

Two check lists are used routinely in the statistical assessment of manuscripts submitted to the *British Medical Journal*. One is for papers of a general nature and the other specifically for reports on clinical trials. Each check list includes questions on the design, conduct, analysis, and presentation of studies, and answers to these contribute to the overall statistical evaluation.

Only a small proportion of submitted papers are assessed statistically, and these are selected at the refereeing or editorial stage. Examination of the use of the check lists showed that most papers contained statistical failings, many of which could easily be remedied.

It is recommended that the check lists should be used by statistical referees, editorial staff, and authors and also during the design stage of studies.

Introduction

The *British Medical Journal* uses two check lists to evaluate the statistical aspects of medical studies. These check lists have been developed during statistical assessment of papers submitted to the journal[1] and have been influenced by others published previously.[2–5] One check list is intended for all studies other than clinical trials and, because of this non-specific application, is limited in detail. The second is for clinical trials and includes questions concerned with randomised or non-randomised treatment or intervention comparisons. Information on the principles behind the questions may be found, for example, in the above publications[1–5] or in the statistical guidelines of chapter 9.

Uses of the check lists

The check lists may be used at different stages of manuscript assessment and study development.

Refereeing is difficult and time consuming,[6–9] but submitted papers clearly require subject matter referees to judge their merit within the medical speciality. Many reports, however, have some statistical content which may be outside the expertise of these particular referees and warrant separate assessment. Although the relevant considerations for this may be clear in a statistician's mind, a list of items to check and respond to serves as a useful reminder. These answers serve as the backbone for the statistician's recommendations on the paper and are supplemented usually with written comments.

Editorial staff find a check list helpful in obtaining a summary view on a paper. Because of the fixed format they can develop a familiarity which allows more rapid evaluation than from a textual report. The latter will generally be needed as well, but will be shorter than a report without the check list.

Authors receiving a copy of the completed check list from the editor can see where their paper was thought to be statistically unsatisfactory—if that is the case. Suggestions for improvements will usually be given in the report if revision is suggested. Alternatively, problems with the design or conduct of the study making the paper unsuitable for publication will be pointed out; some examples are given by Vaisrub.[10]

Planners of studies can be guided by the check lists, which indicate the need to consider relevant statistical aspects during development of protocols. Detailed advice may have to be sought from a statistician or in appropriate publications. Referral to the check lists should also improve the description of the statistical aspects of studies in submitted papers.

Outline of the check lists

General check list

Aspects covered by the general check list include design, conduct, analysis, and presentation of studies (fig 10.1). For each

BMJ Ref No: ______________________ Date of Review: ______________

Design features			
1 Was the objective of the study sufficiently described?	Yes	Unclear	No
2 Was an appropriate study design used to achieve the objective?	Yes	Unclear	No
3 Was there a satisfactory statement given of source of subjects?	Yes	Unclear	No
4 Was a pre-study calculation of required sample size reported?		Yes	No
Conduct of study			
5 Was a satisfactory response rate achieved?	Yes	Unclear	No
Analysis and presentation			
6 Was there a statement adequately describing or referencing all statistical procedures used?		Yes	No
7 Were the statistical analyses used appropriate?	Yes	Unclear	No
8 Was the presentation of statistical material satisfactory?		Yes	No
9 Were confidence intervals given for the main results?		Yes	No
10 Was the conclusion drawn from the statistical analysis justified?	Yes	Unclear	No
Recommendation on paper			
11 Is the paper of acceptable statistical standard for publication?		Yes	No
12 If "No" to Question 11, could it become acceptable with suitable revision?		Yes	No

Reviewer: ______________________

FIG 10.1—Check list for statistical review of general papers for the *British Medical Journal*.

question "yes" or "no" answers are sought, but in some cases "unclear" is allowed, though its use should be minimal.

The first part of the check list relates to considerations before the start of an investigation, such as defining its main objective(s). Sometimes a choice of suitable studies to meet these is available, but some designs will be inappropriate. For example, it would not be sensible to compare elderly diseased patients with young healthy adults to determine whether a blood constituent is aetiologically important. Design considerations also include techniques for measurement and collection of data. In addition, important statistical questions relate to the source and number of subjects studied. The former will be relevant to the validity of any generalised inferences from the results. The issue of the sample size required for a study is well documented, but many studies are still too small to detect anything other than large, and often unrealistic, effects.

When the study is under way a high participation rate is needed from the recruited subjects. Those who do not participate fully are almost certain to be a biased group in some respects, with detrimental effects on the interpretation of the results. A comparison of relevant characteristics of responders and non-responders should be given.

The statistical methods used should be stated. If a technique is novel or unfamiliar then a description of its purpose and an outline of the method should be given together with a suitable reference. Aspects of presentation will also be checked, including tables and figures as well as textual content.

From the answers to the check list a summary can be made of the statistical content of a paper. Other features, which may be mentioned in the accompanying written report, contribute to the recommendation on its statistical quality.

Clinical trials check list

For clinical trials specific questions may be asked in addition to the items from the general check list (fig 10.2).

At the design stage of a clinical trial it is important to determine the diagnostic criteria for inclusion of subjects and to define clearly the treatments to be compared. Where a randomised study is appropriate, which usually is the case, a method of random allocation to treatment is mandatory and should be clearly described. Unambiguous measures of outcome must be specified for trials comparing treatments and the duration of follow up stated. There are advantages if double blind comparisons can be made, and treatment should start with a minimum delay after patient allocation. All these features should be described in the trial protocol.

In the results section the numbers and proportions of subjects treated and followed up should be stated. It is important also to describe drop outs and side effects by treatment group. In addition, treatment groups should be compared for relevant prognostic characteristics and adjustments for these made if appropriate in the analysis of outcome.

Experience so far

We have used the check lists on a regular basis for only a short period, so that a limited amount of descriptive data are available on

BMJ Ref No: ______________________ Date of Review: ______________

Design features				
1	Was the objective of the trial sufficiently described?	Yes	Unclear	No
2	Was there a satisfactory statement given of diagnostic criteria for entry to trial?	Yes	Unclear	No
3	Was there a satisfactory statement given of source of subjects?	Yes	Unclear	No
4	Were concurrent controls used (as opposed to historical controls)?	Yes	Unclear	No
5	Were the treatments well defined?	Yes	Unclear	No
6	Was random allocation to treatment used?	Yes	Unclear	No
7	Was the method of randomisation described?	Yes	Unclear	No
8	Was there an acceptably short delay from allocation to commencement of treatment?	Yes	Unclear	No
9	Was the potential degree of blindness used?	Yes	Unclear	No
10	Was there a satisfactory statement of criteria for outcome measures?	Yes	Unclear	No
11	Were the outcome measures appropriate?	Yes	Unclear	No
12	Was a pre-study calculation of required sample size reported?		Yes	No
13	Was the duration of post-treatment follow up stated?	Yes	Unclear	No
Conduct of trial				
14	Were the treatment and control groups comparable in relevant measures?	Yes	Unclear	No
15	Were a high proportion of the subjects followed up?	Yes	Unclear	No
16	Did a high proportion of subjects complete treatment?	Yes	Unclear	No
17	Were the drop outs described by treatment/control groups?	Yes	Unclear	No
18	Were side effects of treatment reported?	Yes	Unclear	No
Analysis and presentation				
19	Was there a statement adequately describing or referencing all statistical procedures used?		Yes	No
20	Were the statistical analyses used appropriate?	Yes	Unclear	No
21	Were prognostic factors adequately considered?	Yes	Unclear	No
22	Was the presentation of statistical material satisfactory?		Yes	No
23	Were confidence intervals given for the main results?		Yes	No
24	Was the conclusion drawn from the statistical analysis justified?	Yes	Unclear	No
Recommendation on paper				
25	Is the paper of acceptable statistical standard for publication?		Yes	No
26	If "No" to Question 25, could it become acceptable with suitable revision?		Yes	No

Reviewer: ______________________________

FIG 10.2—Check list for statistical review of papers on clinical trials for the *British Medical Journal*.

the main statistical problems found. We do, however, have preliminary findings based on 103 papers for which the general check list was used and 45 papers on clinical trials. Each of these papers was referred for statistical assessment because of comments by the subject matter referee or the editorial staff, and they are a small and unrepresentative sample of papers submitted to (or published in) the *British Medical Journal*. Thus the descriptive figures given below have not been subjected to any formal statistical analysis.

General check list

For the general papers design features were the most satisfactory. Nevertheless, for 28 of the 103 papers the appropriateness of the study design was in doubt, and in 22 papers the source of subjects was not clear. In only one paper did the authors report calculating a required sample size in advance. Response rates were thought to be satisfactory in 84 of the 100 papers where the question was appropriate, but for 12 of the other 16 this information was not clearly given.

In relation to analysis about a third (34) of the papers did not describe the statistical procedures used, and in only 42 papers were the methods said to be appropriate. The main adverse comments related to lack of allowance for confounding variables, invalid use of the χ^2 test, unsuitable analysis of non-Normal data, problems of multiple comparisons, and incorrect arithmetic. Presentation was assessed as unsatisfactory for 76 of the 103 papers. The most frequent difficulties related to problems with tables, inadequate descriptions of the outcomes of hypothesis tests, lack of confidence intervals, non-Normal data, and notational ambiguities mainly associated with use of the $\pm$ sign (now banned by the *British Medical Journal* and other journals—see chapter 2: appendix 1). In only 35 papers was the conclusion drawn from the statistical analysis thought to be justified. Overall as few as 17 of the 103 papers were regarded as statistically acceptable for publication. Only six papers, however, were thought to be unsuitable for revision, though in 40 cases it was "unclear" whether revision was possible.

Clinical trials check list

For the 45 papers on clinical trials the design aspects were again reasonable according to the statistical assessors. The main points of exception were a lack of description of the method of randomisation in 35 papers and the absence of a power based calculation of sample size in 38. The latter raises important ethical as well as statistical considerations,[11] which apply to the general papers also. Questions on the delay between allocation and beginning treatment and on the potential degree of blindness used were answered as "unclear" for 18 and 22 of the papers respectively.

That part of the check list concerned with statistical analysis disclosed a situation similar to that in the general papers. The method was neither described nor referenced in 25 papers and was said to be inappropriate in 19. Prognostic factors were reported to be inadequately considered in 24 papers and presentation as unsatisfactory in 41. The conclusion from the statistical analysis was said to be unjustified or in doubt in 31 of the 45 papers. For only five of the 41 papers considered unacceptable for publication, however, was suitable revision not thought possible—three of them being non-randomised studies.

Comments

These check lists have evolved over a period of time and, as shown in figs 10.1 and 10.2, differ slightly from those used initially. For example, the question on confidence intervals (question 9 in the general check list, question 23 in the clinical trials check list) is a recent addition. It has been included partly as a consequence of a change in *British Medical Journal* policy.[12]

From this preliminary look at answers to the check lists improvements in reporting statistical procedures are clearly needed by some authors. Quite often the problems which have been found relate to easily rectifiable omissions of information, though sometimes there are more serious difficulties in analysis. Statistical assessment is mentioned as a possibility for any article submitted to the *British Medical Journal*[13] and the check lists are now used routinely in this. Such a statistical evaluation is one way to prevent the publication of papers with unsatisfactory statistical content. Other approaches are, of course, possible,[14] including the adoption of published statistical guidelines (such as in chapter 9) or having a statistician on the editorial board, or both. The check lists are intended for guidance on the statistical content of papers and are not presented as items to be covered at the expense of other important aspects of medical studies.[15 16]

1 Gardner MJ, Altman DG, Jones DR, Machin D. Is the statistical assessment of papers submitted to the "British Medical Journal" effective? *Br Med J*. 1983;**286**:1485–8.
2 Lionel NDW, Herxheimer A. Assessing reports of therapeutic trials. *Br Med J* 1970;**iii**:637–40.
3 Ford BL, Tortora RD. A consulting aid to sample design. *Biometrics* 1978;**34**:299–304.
4 Sackett DL. Evaluation: requirements for clinical application. In: Warren KS, ed. *Coping with the biomedical literature: a primer for the scientist and the clinician*. New York: Praeger, 1981:123–57.
5 Bland JM, Jones DR, Bennett S, Cook DG, Haines AP, MacFarlane AJ. Is the clinical trials evidence about new drugs statistically adequate? *Br J Clin Pharmacol* 1985;**19**:155–60.
6 Smith R. Steaming up windows and refereeing medical papers. *Br Med J* 1982;**285**:1259–61.
7 Anonymous. Peer review at work. Dean JW, Fowler PBS. Exaggerated responsiveness to thyrotropin releasing hormone: a risk factor in women with coronary artery disease. *Br Med J* 1985:**290**: 1555–61.
8 Commentators. Peer review at work. *Br Med J* 1985;**290**:1743;**290**:1984–5;**291**:412;**291**:414; **291**:485–6.
9 Lock S. *A difficult balance: editorial peer review in medicine*. London: Nuffield Provincial Hospitals Trust, 1985.
10 Vaisrub N. Manuscript review from a statistician's perspective. *JAMA* 1985;**253**:3145–7.
11 Altman DG. Statistics and ethics in medical research: III—how large a sample? *Br Med J* 1980;**281**:1336–8.
12 Langman MJS. Towards estimation and confidence intervals. *Br Med J* 1986;**292**:716.
13 Instructions to authors. *Br Med J* 1988;**296**:48–9.
14 George SL. Statistics in medical journals: a survey of current policies and proposals for editors. *Med Pediatr Oncol* 1985;**13**:109–12.
15 Jones RS. Statistical assessment of papers submitted to the "British Medical Journal." *Br Med J* 1983;**286**:1971.
16 Healy MJR. Statistical guidelines for contributors to medical journals. *Br Med J* 1983;**287**:132.

Part III
Tables

Tables for the calculation of confidence intervals

MARTIN J GARDNER

In each table selected values are given to enable the calculation of 90%, 95%, and 99% confidence intervals using the methods described in part I. For values applicable to other levels of confidence, reference should be made to more extensive published tables such as in *Geigy Scientific Tables*.[1] The first three tables can be used for calculating confidence intervals for a wide variety of statistics—such as means, proportions, regression analyses, and standardised mortality ratios—whereas the last three are for specific statistics, as described in chapter 8.

Table 1	Normal distribution
Table 2	*t* distribution
Table 3	Poisson distribution
Table 4	Median (single sample) or differences between medians (paired samples), based on Binomial distribution with probability $\frac{1}{2}$
Table 5	Differences between medians (unpaired samples), based on distributions of the Wilcoxon two sample rank sum test statistic and of the Mann–Whitney U test statistic
Table 6	Median (single sample) or differences between medians (paired samples), based on the distribution of the Wilcoxon matched pairs signed rank sum test statistic

The tables have been produced directly from theoretical formulae.

1 Lentner C, ed. *Geigy scientific tables*. Vol. 2. 8th ed. Basle: Ciba-Geigy, 1982.

TABLE 1—Values from the Normal distribution for use in calculating confidence intervals

The value tabulated is $N_{1-\alpha/2}$ *from the standard Normal distribution for the* $100(1-\alpha/2)$ *percentile and is to be used in finding* $100(1-\alpha)\%$ *confidence intervals. For a 90% confidence interval* α *is* 0·10, *for a 95% confidence interval* α *is* 0·05, *and for a 99% confidence interval* α *is* 0·01.

Level of confidence		
90%	95%	99%
1·645	1·960	2·576

TABLE 2—Values from the t distribution for 1 to 400 degrees of freedom for use in calculating confidence intervals

The value tabulated is $t_{1-\alpha/2}$ *from the t distribution for the* $100(1-\alpha/2)$ *percentile and is to be used in finding* $100(1-\alpha)\%$ *confidence intervals. For a 90% confidence interval* α *is* 0·10, *for a 95% confidence interval* α *is* 0·05, *and for a 99% confidence interval* α *is* 0·01. *The relation of the degrees of freedom to sample size(s) depends on the particular application and is described in chapters* 2, 3, *and* 5 *where appropriate.*

Degrees of freedom	Level of confidence			Degrees of freedom	Level of confidence		
	90%	95%	99%		90%	95%	99%
1	6·314	12·706	63·657	45	1·679	2·014	2·690
2	2·920	4·303	9·925	46	1·679	2·013	2·687
3	2·353	3·182	5·841	47	1·678	2·012	2·685
4	2·132	2·776	4·604	48	1·677	2·011	2·682
5	2·015	2·571	4·032	49	1·677	2·010	2·680
6	1·943	2·447	3·707	50	1·676	2·009	2·678
7	1·895	2·365	3·499	51	1·675	2·008	2·676
8	1·860	2·306	3·355	52	1·675	2·007	2·674
9	1·833	2·262	3·250	53	1·674	2·006	2·672
10	1·812	2·228	3·169	54	1·674	2·005	2·670
11	1·796	2·201	3·106	55	1·673	2·004	2·668
12	1·782	2·179	3·055	56	1·673	2·003	2·667
13	1·771	2·160	3·012	57	1·672	2·002	2·665
14	1·761	2·145	2·977	58	1·672	2·002	2·663
15	1·753	2·131	2·947	59	1·671	2·001	2·662
16	1·746	2·120	2·921	60	1·671	2·000	2·660
17	1·740	2·110	2·898	61	1·670	2·000	2·659
18	1·734	2·101	2·878	62	1·670	1·999	2·657
19	1·729	2·093	2·861	63	1·669	1·998	2·656
20	1·725	2·086	2·845	64	1·669	1·998	2·655
21	1·721	2·080	2·831	65	1·669	1·997	2·654
22	1·717	2·074	2·819	66	1·668	1·997	2·652
23	1·714	2·069	2·807	67	1·668	1·996	2·651
24	1·711	2·064	2·797	68	1·668	1·995	2·650
25	1·708	2·060	2·787	69	1·667	1·995	2·649
26	1·706	2·056	2·779	70	1·667	1·994	2·648
27	1·703	2·052	2·771	71	1·667	1·994	2·647
28	1·701	2·048	2·763	72	1·666	1·993	2·646
29	1·699	2·045	2·756	73	1·666	1·993	2·645
30	1·697	2·042	2·750	74	1·666	1·993	2·644
31	1·696	2·040	2·744	75	1·665	1·992	2·643
32	1·694	2·037	2·738	76	1·665	1·992	2·642
33	1·692	2·035	2·733	77	1·665	1·991	2·641
34	1·691	2·032	2·728	78	1·665	1·991	2·640
35	1·690	2·030	2·724	79	1·664	1·990	2·640
36	1·688	2·028	2·719	80	1·664	1·990	2·639
37	1·687	2·026	2·715	81	1·664	1·990	2·638
38	1·686	2·024	2·712	82	1·664	1·989	2·637
39	1·685	2·023	2·708	83	1·663	1·989	2·636
40	1·684	2·021	2·704	84	1·663	1·989	2·636
41	1·683	2·020	2·701	85	1·663	1·988	2·635
42	1·682	2·018	2·698	86	1·663	1·988	2·634
43	1·681	2·017	2·695	87	1·663	1·988	2·634
44	1·680	2·015	2·692	88	1·662	1·987	2·633

TABLE 2—(*continued*)

Degrees of freedom	Level of confidence 90%	95%	99%
89	1·662	1·987	2·632
90	1·662	1·987	2·632
91	1·662	1·986	2·631
92	1·662	1·986	2·630
93	1·661	1·986	2·630
94	1·661	1·986	2·629
95	1·661	1·985	2·629
96	1·661	1·985	2·628
97	1·661	1·985	2·627
98	1·661	1·984	2·627
99	1·660	1·984	2·626
100	1·660	1·984	2·626
101	1·660	1·984	2·625
102	1·660	1·983	2·625
103	1·660	1·983	2·624
104	1·660	1·983	2·624
105	1·660	1·983	2·623
106	1·659	1·983	2·623
107	1·659	1·982	2·623
108	1·659	1·982	2·622
109	1·659	1·982	2·622
110	1·659	1·982	2·621
111	1.659	1·982	2·621
112	1.659	1·981	2·620
113	1.658	1·981	2·620
114	1.658	1·981	2·620
115	1.658	1·981	2·619
116	1.658	1·981	2·619
117	1.658	1·980	2·619
118	1.658	1·980	2·618
119	1.658	1·980	2·618
120	1.658	1·980	2·617
121	1.658	1·980	2·617
122	1.657	1·980	2·617
123	1.657	1·979	2·616
124	1.657	1·979	2·616
125	1.657	1·979	2·616
126	1.657	1·979	2·615
127	1.657	1·979	2·615
128	1.657	1·979	2·615
129	1.657	1·979	2·614
130	1.657	1·978	2·614
131	1.657	1·978	2·614
132	1.656	1·978	2·614
133	1.656	1·978	2·613
134	1.656	1·978	2·613
135	1.656	1·978	2·613
136	1.656	1·978	2·612
137	1.656	1·977	2·612
138	1.656	1·977	2·612
139	1.656	1·977	2·612
140	1.656	1·977	2·611
141	1.656	1·977	2·611
142	1.656	1·977	2·611
143	1.656	1·977	2·611
144	1.656	1·977	2·610
145	1.655	1·976	2·610
146	1.655	1·976	2·610
147	1.655	1·976	2·610
148	1.655	1·976	2·609
149	1.655	1·976	2·609
150	1.655	1·976	2·609
151	1.655	1·976	2·609
152	1.655	1·976	2·609
153	1.655	1·976	2·608
154	1.655	1·975	2·608
155	1.655	1·975	2·608
156	1.655	1·975	2·608
157	1.655	1·975	2·608
158	1.655	1·975	2·607
159	1.654	1·975	2·607
160	1.654	1·975	2·607
161	1·654	1·975	2·607
162	1·654	1·975	2·607
163	1·654	1·975	2·606
164	1·654	1·975	2·606
165	1·654	1·974	2·606
166	1·654	1·974	2·606
167	1·654	1·974	2·606
168	1·654	1·974	2·605
169	1·654	1·974	2·605
170	1·654	1·974	2·605
171	1·654	1·974	2·605
172	1·654	1·974	2·605
173	1·654	1·974	2·605
174	1·654	1·974	2·604
175	1·654	1·974	2·604
176	1·654	1·974	2·604
177	1·654	1·973	2·604
178	1·653	1·973	2·604
179	1·653	1·973	2·604
180	1·653	1·973	2·603
181	1·653	1·973	2·603
182	1·653	1·973	2·603
183	1·653	1·973	2·603
184	1·653	1·973	2·603
185	1·653	1·973	2·603
186	1·653	1·973	2·603
187	1·653	1·973	2·602
188	1·653	1·973	2·602
189	1·653	1·973	2·602
190	1·653	1·973	2·602

TABLE 2—(*continued*)

Degrees of freedom	Level of confidence			Degrees of freedom	Level of confidence		
	90%	95%	99%		90%	95%	99%
191	1·653	1·972	2·602	270	1·651	1·969	2·594
192	1·653	1·972	2·602	280	1·650	1·968	2·594
193	1·653	1·972	2·602	290	1·650	1·968	2·593
194	1·653	1·972	2·601	300	1·650	1·968	2·592
195	1·653	1·972	2·601	310	1·650	1·968	2·592
196	1·653	1·972	2·601	320	1·650	1·967	2·591
197	1·653	1·972	2·601	330	1·649	1·967	2·591
198	1·653	1·972	2·601	340	1·649	1·967	2·590
199	1·653	1·972	2·601	350	1·649	1·967	2·590
200	1·653	1·972	2·601	360	1·649	1·967	2·590
210	1·652	1·971	2·599	370	1·649	1·966	2·589
220	1·652	1·971	2·598	380	1·649	1·966	2·589
230	1·652	1·970	2·597	390	1·649	1·966	2·588
240	1·651	1·970	2·596	400	1·649	1·966	2·588
250	1·651	1·969	2·596	∞	1·645	1·960	2·576
260	1·651	1·969	2·595				

TABLE 3—Values from the Poisson distribution for observed numbers of from 0 to 100 for use in calculating confidence intervals

If x is the observed number in the study then the values tabulated (x_L to x_U) give the 100(1 − α)% confidence interval for the population mean, assuming that the observed number is from a Poisson distribution. For a 90% confidence interval α is 0·10, for a 95% confidence interval α is 0·05, and for a 99% confidence interval α is 0·01.

	Level of confidence					
	90%		95%		99%	
x	x_L	x_U	x_L	x_U	x_L	x_U
0	0	2·996	0	3·689	0	5·298
1	0·051	4·744	0·025	5·572	0·005	7·430
2	0·355	6·296	0·242	7·225	0·103	9·274
3	0·818	7·754	0·619	8·767	0·338	10·977
4	1·366	9·154	1·090	10·242	0·672	12·594
5	1·970	10·513	1·623	11·668	1·078	14·150
6	2·613	11·842	2·202	13·059	1·537	15·660
7	3·285	13·148	2·814	14·423	2·037	17·134
8	3·981	14·435	3·454	15·763	2·571	18·578
9	4·695	15·705	4·115	17·085	3·132	19·998
10	5·425	16·962	4·795	18·390	3·717	21·398
11	6·169	18·208	5·491	19·682	4·321	22·779
12	6·924	19·443	6·201	20·962	4·943	24·145
13	7·690	20·669	6·922	22·230	5·580	25·497
14	8·464	21·886	7·654	23·490	6·231	26·836
15	9·246	23·097	8·395	24·740	6·893	28·164
16	10·036	24·301	9·145	25·983	7·567	29·482
17	10·832	25·499	9·903	27·219	8·251	30·791
18	11·634	26·692	10·668	28·448	8·943	32·091
19	12·442	27·879	11·439	29·671	9·644	33·383
20	13·255	29·062	12·217	30·888	10·353	34·668
21	14·072	30·240	12·999	32·101	11·069	35·946
22	14·894	31·415	13·787	33·308	11·792	37·218
23	15·719	32·585	14·580	34·511	12·521	38·484
24	16·549	33·752	15·377	35·710	13·255	39·745
25	17·382	34·916	16·179	36·905	13·995	41·000
26	18·219	36·077	16·984	38·096	14·741	42·251
27	19·058	37·234	17·793	39·284	15·491	43·497
28	19·901	38·389	18·606	40·468	16·245	44·738
29	20·746	39·541	19·422	41·649	17·004	45·976
30	21·594	40·691	20·241	42·827	17·767	47·209
31	22·445	41·838	21·063	44·002	18·534	48·439
32	23·297	42·982	21·888	45·174	19·305	49·665
33	24·153	44·125	22·716	46·344	20·079	50·888
34	25·010	45·266	23·546	47·512	20·857	52·107
35	25·870	46·404	24·379	48·677	21·638	53·324
36	26·731	47·541	25·214	49·839	22·422	54·537
37	27·595	48·675	26·051	51·000	23·208	55·748
38	28·460	49·808	26·891	52·158	23·998	56·955
39	29·327	50·940	27·733	53·314	24·791	58·161
40	30·196	52·069	28·577	54·469	25·586	59·363
41	31·066	53·197	29·422	55·621	26·384	60·563
42	31·938	54·324	30·270	56·772	27·184	61·761

TABLE 3—(*continued*)

	Level of confidence					
	90%		95%		99%	
x	x_L	x_U	x_L	x_U	x_L	x_U
43	32·812	55·449	31·119	57·921	27·986	62·956
44	33·687	56·573	31·970	59·068	28·791	64·149
45	34·563	57·695	32·823	60·214	29·598	65·341
46	35·441	58·816	33·678	61·358	30·407	66·530
47	36·320	59·935	34·534	62·500	31·218	67·717
48	37·200	61·054	35·391	63·641	32·032	68·902
49	38·082	62·171	36·250	64·781	32·847	70·085
50	38·965	63·287	37·111	65·919	33·664	71·266
51	39·849	64·402	37·973	67·056	34·483	72·446
52	40·734	65·516	38·836	68·191	35·303	73·624
53	41·620	66·628	39·701	69·325	36·125	74·800
54	42·507	67·740	40·566	70·458	36·949	75·974
55	43·396	68·851	41·434	71·590	37·775	77·147
56	44·285	69·960	42·302	72·721	38·602	78·319
57	45·176	71·069	43·171	73·850	39·431	79·489
58	46·067	72·177	44·042	74·978	40·261	80·657
59	46·959	73·284	44·914	76·106	41·093	81·824
60	47·852	74·390	45·786	77·232	41·926	82·990
61	48·746	75·495	46·660	78·357	42·760	84·154
62	49·641	76·599	47·535	79·481	43·596	85·317
63	50·537	77·702	48·411	80·604	44·433	86·479
64	51·434	78·805	49·288	81·727	45·272	87·639
65	52·331	79·907	50·166	82·848	46·111	88·798
66	53·229	81·008	51·044	83·968	46·952	89·956
67	54·128	82·108	51·924	85·088	47·794	91·112
68	55·028	83·208	52·805	86·206	48·637	92·269
69	55·928	84·306	53·686	87·324	49·482	93·423
70	56·830	85·405	54·568	88·441	50·327	94·577
71	57·732	86·502	55·452	89·557	51·174	95·729
72	58·634	87·599	56·336	90·672	52·022	96·881
73	59·537	88·695	57·220	91·787	52·871	98·031
74	60·441	89·790	58·106	92·900	53·720	99·180
75	61·346	90·885	58·992	94·013	54·571	100·328
76	62·251	91·979	59·879	95·125	55·423	101·476
77	63·157	93·073	60·767	96·237	56·276	102·622
78	64·063	94·166	61·656	97·348	57·129	103·767
79	64·970	95·258	62·545	98·458	57·984	104·912
80	65·878	96·350	63·435	99·567	58·840	106·056
81	66·786	97·441	64·326	100·676	59·696	107·198
82	67·965	98·532	65·217	101·784	60·553	108·340
83	68·604	99·622	66·109	102·891	61·412	109·481
84	69·514	100·712	67·002	103·998	62·271	110·621
85	70·425	101·801	67·895	105·104	63·131	111·761
86	71·336	102·889	68·789	106·209	63·991	112·899
87	72·247	103·977	69·683	107·314	64·853	114·037
88	73·159	105·065	70·579	108·418	65·715	115·174
89	74·071	106·152	71·474	109·522	66·578	116·310
90	74·984	107·239	72·371	110·625	67·442	117·445
91	75·898	108·325	73·268	111·728	68·307	118·580

TABLE 3—*(continued)*

	Level of confidence					
	90%		95%		99%	
x	x_L	x_U	x_L	x_U	x_L	x_U
92	76·812	109·410	74·165	112·830	69·172	119·714
93	77·726	110·495	75·063	113·931	70·038	120·847
94	78·641	111·580	75·962	115·032	70·905	121·980
95	79·556	112·664	76·861	116·133	71·773	123·112
96	80·472	113·748	77·760	117·232	72·641	124·243
97	81·388	114·832	78·660	118·332	73·510	125·373
98	82·305	115·915	79·561	119·431	74·379	126·503
99	83·222	116·997	80·462	120·529	75·250	127·632
100	84·139	118·079	81·364	121·627	76·120	128·761

For $x > 100$ the following calculations can be carried out to obtain approximate values for x_L and x_U:

$$x_L = \left(\frac{N_{1-\alpha/2}}{2} - \sqrt{x}\right)^2 \quad \text{and} \quad x_U = \left(\frac{N_{1-\alpha/2}}{2} + \sqrt{x+1}\right)^2,$$

where $N_{1-\alpha/2}$ is the appropriate value from the standard Normal distribution for the 100 $(1-\alpha/2)$ percentile.

As an example of the closeness of the approximations to the exact values, for $\alpha = 0{\cdot}05$ ($N_{1-\alpha/2} = 1{\cdot}96$) and $x = 100$ the formulae give $x_L = 81{\cdot}360$ and $x_U = 121{\cdot}658$ for the 95% confidence interval compared to the values of $x_L = 81{\cdot}364$ and $x_U = 121{\cdot}627$ tabulated above.

TABLE 4—Ranks of the observations for use in calculating confidence intervals for population medians in single samples or for differences between population medians for the case of two paired samples with sample sizes from 6 to 100 and the associated exact levels of confidence, based on the Binomial distribution with probability $\frac{1}{2}$

The values tabulated (r_L to r_U) show the ranks of the observations to be used to give the approximate $100(1-\alpha)\%$ *confidence interval for the population median. For a 90% confidence interval α is 0·10, for a 95% confidence interval α is 0·05, and for a 99% confidence interval α is 0·01.*

Sample size (n)	Level of confidence								
	90% (approx)			95% (approx)			99% (approx)		
	r_L	r_U	Exact level (%)	r_L	r_U	Exact level (%)	r_L	r_U	Exact level (%)
6	1	6	96·9	1	6	96·9	—	—	—
7	1	7	98·4	1	7	98·4	—	—	—
8	2	7	93·0	1	8	99·2	1	8	99·2
9	2	8	96·1	2	8	96·1	1	9	99·6
10	2	9	97·9	2	9	97·9	1	10	99·8
11	3	9	93·5	2	10	98·8	1	11	99·9
12	3	10	96·1	3	10	96·1	2	11	99·4
13	4	10	90·8	3	11	97·8	2	12	99·7
14	4	11	94·3	3	12	98·7	2	13	99·8
15	4	12	96·5	4	12	96·5	3	13	99·3
16	5	12	92·3	4	13	97·9	3	14	99·6
17	5	13	95·1	5	13	95·1	3	15	99·8
18	6	13	90·4	5	14	96·9	4	15	99·2
19	6	14	93·6	5	15	98·1	4	16	99·6
20	6	15	95·9	6	15	95·9	4	17	99·7
21	7	15	92·2	6	16	97·3	5	17	99·3
22	7	16	94·8	6	17	98·3	5	18	99·6
23	8	16	90·7	7	17	96·5	5	19	99·7
24	8	17	93·6	7	18	97·7	6	19	99·3
25	8	18	95·7	8	18	95·7	6	20	99·6
26	9	18	92·4	8	19	97·1	7	20	99·1
27	9	19	94·8	8	20	98·1	7	21	99·4
28	10	19	91·3	9	20	96·4	7	22	99·6
29	10	20	93·9	9	21	97·6	8	22	99·2
30	11	20	90·1	10	21	95·7	8	23	99·5
31	11	21	92·9	10	22	97·1	8	24	99·7
32	11	22	95·0	10	23	98·0	9	24	99·3
33	12	22	92·0	11	23	96·5	9	25	99·5
34	12	23	94·2	11	24	97·6	10	25	99·1
35	13	23	91·0	12	24	95·9	10	26	99·4
36	13	24	93·5	12	25	97·1	10	27	99·6
37	14	24	90·1	13	25	95·3	11	27	99·2
38	14	25	92·7	13	26	96·6	11	28	99·5
39	14	26	94·7	13	27	97·6	12	28	99·1
40	15	26	91·9	14	27	96·2	12	29	99·4
41	15	27	94·0	14	28	97·2	12	30	99·6
42	16	27	91·2	15	28	95·6	13	30	99·2
43	16	28	93·4	15	29	96·8	13	31	99·5
44	17	28	90·4	16	29	95·1	14	31	99·0
45	17	29	92·8	16	30	96·4	14	32	99·3

TABLE 4—(*continued*)

Sample size (n)	Level of confidence								
	90% (approx)			95% (approx)			99% (approx)		
	r_L	r_U	Exact level (%)	r_L	r_U	Exact level (%)	r_L	r_U	Exact level (%)
46	17	30	94·6	16	31	97·4	14	33	99·5
47	18	30	92·1	17	31	96·0	15	33	99·2
48	18	31	94·1	17	32	97·1	15	34	99·4
49	19	31	91·5	18	32	95·6	16	34	99·1
50	19	32	93·5	18	33	96·7	16	35	99·3
51	20	32	90·8	19	33	95·1	16	36	99·5
52	20	33	93·0	19	34	96·4	17	36	99·2
53	21	33	90·2	19	35	97·3	17	37	99·5
54	21	34	92·4	20	35	96·0	18	37	99·1
55	21	35	94·2	20	36	97·0	18	38	99·4
56	22	35	91·9	21	36	95·6	18	39	99·5
57	22	36	93·7	21	37	96·7	19	39	99·2
58	23	36	91·3	22	37	95·2	19	40	99·5
59	23	37	93·3	22	38	96·4	20	40	99·1
60	24	37	90·8	22	39	97·3	20	41	99·4
61	24	38	92·8	23	39	96·0	21	41	99·0
62	25	38	90·2	23	40	97·0	21	42	99·3
63	25	39	92·3	24	40	95·7	21	43	99·5
64	25	40	94·0	24	41	96·7	22	43	99·2
65	26	40	91·8	25	41	95·4	22	44	99·4
66	26	41	93·6	25	42	96·4	23	44	99·1
67	27	41	91·4	26	42	95·0	23	45	99·3
68	27	42	93·2	26	43	96·2	23	46	99·5
69	28	42	90·9	26	44	97·1	24	46	99·2
70	28	43	92·8	27	44	95·9	24	47	99·4
71	29	43	90·4	27	45	96·8	25	47	99·1
72	29	44	92·4	28	45	95·6	25	48	99·4
73	29	45	94.0	28	46	96·6	26	48	99·0
74	30	45	91·9	29	46	95·3	26	49	99·3
75	30	46	93·6	29	47	96·3	26	50	99·5
76	31	46	91·5	29	48	97·1	27	50	99·2
77	31	47	93·2	30	48	96·0	27	51	99·4
78	32	47	91·1	30	49	96·9	28	51	99·1
79	32	48	92·9	31	49	95·8	28	52	99·3
80	33	48	90·7	31	50	96·7	29	52	99·0
81	33	49	92·5	32	50	95·5	29	53	99·3
82	34	49	90·3	32	51	96·5	29	54	99·5
83	34	50	92·2	33	51	95·2	30	54	99·2
84	34	51	93·7	33	52	96·2	30	55	99·4
85	35	51	91·8	33	53	97·1	31	55	99·1
86	35	52	93·4	34	53	96·0	31	56	99·3
87	36	52	91·4	34	54	96·9	32	56	99·0
88	36	53	93·1	35	54	95·8	32	57	99·3
89	37	53	91·1	35	55	96·7	32	58	99·4
90	37	54	92·8	36	55	95·5	33	58	99·2
91	38	54	90·7	36	56	96·5	33	59	99·4
92	38	55	92·4	37	56	95·3	34	59	99·1
93	39	55	90·3	37	57	96·2	34	60	99·3

TABLE 4—*(continued)*

Sample size (n)	Level of confidence								
	90% (approx)			95% (approx)			99% (approx)		
	r_L	r_U	Exact level (%)	r_L	r_U	Exact level (%)	r_L	r_U	Exact level (%)
94	39	56	92·1	38	57	95·1	35	60	99·0
95	39	57	93·6	38	58	96·0	35	61	99·3
96	40	57	91·8	38	59	96·8	35	62	99·4
97	40	58	93·3	39	59	95·8	36	62	99·2
98	41	58	91·5	39	60	96·7	36	63	99·4
99	41	59	93·0	40	60	95·6	37	63	99·1
100	42	59	91·1	40	61	96·5	37	64	99·3

For sample sizes of n over 100, satisfactory approximations to the values of r_L and r_U can be found as described in chapter 8.

For $n = 100$ and $\alpha = 0{\cdot}01$, for example, the calculations give $r = 37{\cdot}1$ and $s = 63{\cdot}9$, which rounded to the nearest integer give $r_L = 37$ and $r_U = 64$ for finding the 99% confidence interval, the same values as shown in the table.

For an explanation of the use of this table see chapter 8.

TABLE 5—Values of K for use in calculating confidence intervals for differences between population medians for the case of two unpaired samples with sample sizes n_1 and n_2 from 5 to 25 and the associated exact levels of confidence, based on the Wilcoxon two sample rank sum distribution

Sample sizes (n_1, n_2)		Level of confidence					
		90% (approx)		95% (approx)		99% (approx)	
Smaller	Larger	K	Exact level (%)	K	Exact level (%)	K	Exact level (%)
5	5	5	90·5	3	96·8	1	99·2
5	6	6	91·8	4	97·0	2	99·1
5	7	7	92·7	6	95·2	2	99·5
5	8	9	90·7	7	95·5	3	99·4
5	9	10	91·7	8	95·8	4	99·3
5	10	12	90·1	9	96·0	5	99·2
5	11	13	91·0	10	96·2	6	99·1
5	12	14	91·8	12	95·2	7	99·1
5	13	16	90·5	13	95·4	8	99·0
5	14	17	91·3	14	95·6	8	99·3
5	15	19	90·2	15	95·8	9	99·2
5	16	20	90·9	16	96·0	10	99·2
5	17	21	91·5	18	95·2	11	99·1
5	18	23	90·6	19	95·4	12	99·1
5	19	24	91·2	20	95·6	13	99·1
5	20	26	90·3	21	95·8	14	99·0
5	21	27	90·9	23	95·1	15	99·0
5	22	29	90·1	24	95·3	15	99·2
5	23	30	90·6	25	95·5	16	99·2
5	24	31	91·1	26	95·6	17	99·1
5	25	33	90·4	28	95·1	18	99·1
6	6	8	90·7	6	95·9	3	99·1
6	7	9	92·7	7	96·5	4	99·2
6	8	11	91·9	9	95·7	5	99·2
6	9	13	91·2	11	95·0	6	99·2
6	10	15	90·7	12	95·8	7	99·3
6	11	17	90·2	14	95·2	8	99·3
6	12	18	91·7	15	95·9	10	99·0
6	13	20	91·3	17	95·4	11	99·1
6	14	22	90·9	18	95·9	12	99·1
6	15	24	90·5	20	95·5	13	99·2
6	16	26	90·2	22	95·1	14	99·2
6	17	27	91·4	23	95·6	16	99·0
6	18	29	91·0	25	95·3	17	99·1
6	19	31	90·8	26	95·7	18	99·1
6	20	33	90·5	28	95·4	19	99·1
6	21	35	90·3	30	95·1	20	99·2
6	22	37	90·0	31	95·5	22	99·0
6	23	38	91·0	33	95·3	23	99·1
6	24	40	90·7	34	95·6	24	99·1
6	25	42	90·5	36	95·4	25	99·1
7	7	12	90·3	9	96·2	5	99·3
7	8	14	90·6	11	96·0	7	99·1
7	9	16	90·9	13	95·8	8	99·2

TABLE 5—*(continued)*

Sample sizes (n_1, n_2)		Level of confidence					
		90% (approx)		95% (approx)		99% (approx)	
Smaller	Larger	K	Exact level (%)	K	Exact level (%)	K	Exact level (%)
7	10	18	91·2	15	95·7	10	99·0
7	11	20	91·5	17	95·6	11	99·2
7	12	22	91·7	19	95·5	13	99·0
7	13	25	90·3	21	95·4	14	99·2
7	14	27	90·6	23	95·4	16	99·0
7	15	29	90·9	25	95·3	17	99·1
7	16	31	91·1	27	95·3	19	99·0
7	17	34	90·1	29	95·3	20	99·1
7	18	36	90·3	31	95·3	22	99·1
7	19	38	90·6	33	95·2	23	99·2
7	20	40	90·8	35	95·2	25	99·1
7	21	42	91·0	37	95·2	26	99·2
7	22	45	90·2	39	95·2	28	99·1
7	23	47	90·4	41	95·2	30	99·0
7	24	49	90·6	43	95·2	31	99·1
7	25	51	90·8	45	95·2	33	99·0
8	8	16	91·7	14	95·0	8	99·3
8	9	19	90·7	16	95·4	10	99·2
8	10	21	91·7	18	95·7	12	99·1
8	11	24	90·9	20	95·9	14	99·1
8	12	27	90·2	23	95·3	16	99·0
8	13	29	91·1	25	95·5	18	99·0
8	14	32	90·5	27	95·8	19	99·2
8	15	34	91·3	30	95·3	21	99·2
8	16	37	90·7	32	95·5	23	99·1
8	17	40	90·3	35	95·1	25	99·1
8	18	42	91·0	37	95·3	27	99·1
8	19	45	90·5	39	95·5	29	99·1
8	20	48	90·1	42	95·1	31	99·0
8	21	50	90·7	44	95·3	33	99·0
8	22	53	90·3	46	95·5	35	99·0
8	23	55	90·9	49	95·2	36	99·1
8	24	58	90·6	51	95·4	38	99·1
8	25	61	90·2	54	95·1	40	99·1
9	9	22	90·6	18	96·0	12	99·2
9	10	25	90·5	21	95·7	14	99·2
9	11	28	90·5	24	95·4	17	99·0
9	12	31	90·5	27	95·1	19	99·1
9	13	34	90·4	29	95·7	21	99·1
9	14	37	90·4	32	95·4	23	99·1
9	15	40	90·4	35	95·2	25	99·2
9	16	43	90·5	38	95·1	28	99·0
9	17	46	90·5	40	95·5	30	99·1
9	18	49	90·5	43	95·4	32	99·1
9	19	52	90·5	46	95·2	34	99·1
9	20	55	90·5	49	95·1	37	99·0
9	21	58	90·6	51	95·5	39	99·1
9	22	61	90·6	54	95·4	41	99·1
9	23	64	90·6	57	95·3	44	99·0

TABLE 5—(*continued*)

Sample sizes (n_1, n_2)		Level of confidence					
		90% (approx)		95% (approx)		99% (approx)	
Smaller	Larger	K	Exact level (%)	K	Exact level (%)	K	Exact level (%)
9	24	67	90·6	60	95·1	46	99·0
9	25	70	90·6	63	95·0	48	99·1
10	10	28	91·1	24	95·7	17	99·1
10	11	32	90·1	27	95·7	19	99·2
10	12	35	90·7	30	95·7	22	99·1
10	13	38	91·2	34	95·1	25	99·0
10	14	42	90·4	37	95·2	27	99·1
10	15	45	90·9	40	95·2	30	99·0
10	16	49	90·3	43	95·3	32	99·1
10	17	52	90·7	46	95·4	35	99·1
10	18	56	90·1	49	95·5	38	99·0
10	19	59	90·6	53	95·0	40	99·1
10	20	63	90·0	56	95·1	43	99·0
10	21	66	90·4	59	95·2	45	99·1
10	22	69	90·8	62	95·3	48	99·1
10	23	73	90·4	65	95·3	51	99·0
10	24	76	90·7	68	95·4	53	99·1
10	25	80	90·3	72	95·0	56	99·1
11	11	35	91·2	31	95·3	22	99·2
11	12	39	90·9	34	95·6	25	99·1
11	13	43	90·7	38	95·3	28	99·1
11	14	47	90·5	41	95·6	31	99·1
11	15	51	90·3	45	95·3	34	99·1
11	16	55	90·1	48	95·6	37	99·1
11	17	58	90·9	52	95·3	40	99·1
11	18	62	90·8	56	95·1	43	99·1
11	19	66	90·6	59	95·3	46	99·1
11	20	70	90·5	63	95·1	49	99·1
11	21	74	90·4	66	95·4	52	99·1
11	22	78	90·3	70	95·2	55	99·1
11	23	82	90·2	74	95·0	58	99·1
11	24	86	90·1	77	95·3	61	99·1
11	25	90	90·0	81	95·1	64	99·1
12	12	43	91·1	38	95·5	28	99·2
12	13	48	90·2	42	95·4	32	99·0
12	14	52	90·5	46	95·4	35	99·1
12	15	56	90·7	50	95·3	38	99·1
12	16	61	90·0	54	95·3	42	99·0
12	17	65	90·3	58	95·2	45	99·1
12	18	69	90·5	62	95·2	48	99·1
12	19	73	90·7	66	95·2	52	99·0
12	20	78	90·1	70	95·2	55	99·1
12	21	82	90·4	74	95·2	59	99·0
12	22	86	90·6	78	95·2	62	99·0
12	23	91	90·1	82	95·1	65	99·1
12	24	95	90·3	86	95·1	69	99·0
12	25	99	90·5	90	95·1	72	99·1

TABLE 5—(*continued*)

Sample sizes (n_1, n_2)		Level of confidence					
		90% (approx)		95% (approx)		99% (approx)	
Smaller	Larger	K	Exact level (%)	K	Exact level (%)	K	Exact level (%)
13	13	52	90·9	46	95·6	35	99·1
13	14	57	90·6	51	95·2	39	99·1
13	15	62	90·2	55	95·4	43	99·0
13	16	66	90·8	60	95·0	46	99·1
13	17	71	90·6	64	95·2	50	99·1
13	18	76	90·3	68	95·4	54	99·0
13	19	81	90·1	73	95·1	58	99·0
13	20	85	90·6	77	95·2	61	99·1
13	21	90	90·4	81	95·4	65	99·1
13	22	95	90·2	86	95·1	69	99·0
13	23	99	90·7	90	95·3	73	99·0
13	24	104	90·5	95	95·1	76	99·1
13	25	109	90·3	99	95·2	80	99·1
14	14	62	90·6	56	95·0	43	99·1
14	15	67	90·7	60	95·4	47	99·1
14	16	72	90·7	65	95·3	51	99·1
14	17	78	90·0	70	95·2	55	99·1
14	18	83	90·1	75	95·1	59	99·1
14	19	88	90·2	79	95·4	64	99·0
14	20	93	90·3	84	95·3	68	99·0
14	21	98	90·4	89	95·2	72	99·0
14	22	103	90·5	94	95·1	76	99·0
14	23	108	90·6	99	95·1	80	99·1
14	24	114	90·0	103	95·3	84	99·1
14	25	119	90·2	108	95·3	88	99·1
15	15	73	90·2	65	95·5	52	99·0
15	16	78	90·7	71	95·1	56	99·1
15	17	84	90·3	76	95·1	61	99·0
15	18	89	90·7	81	95·2	65	99·1
15	19	95	90·4	86	95·3	70	99·0
15	20	101	90·1	91	95·4	74	99·1
15	21	106	90·4	97	95·1	79	99·0
15	22	112	90·2	102	95·1	83	99·1
15	23	117	90·5	107	95·2	88	99·0
15	24	123	90·3	112	95·3	92	99·1
15	25	129	90·0	118	95·0	97	99·0
16	16	84	90·6	76	95·3	61	99·0
16	17	90	90·6	82	95·1	66	99·0
16	18	96	90·5	87	95·4	71	99·0
16	19	102	90·5	93	95·2	75	99·1
16	20	108	90·5	99	95·1	80	99·1
16	21	114	90·5	104	95·3	85	99·1
16	22	120	90·5	110	95·2	90	99·1
16	23	126	90·5	116	95·0	95	99·1
16	24	132	90·5	121	95·2	100	99·1
16	25	138	90·5	127	95·1	105	99·0
17	17	97	90·1	88	95·1	71	99·1

TABLE 5—(*continued*)

Sample sizes (n_1, n_2)		Level of confidence					
		90% (approx)		95% (approx)		99% (approx)	
Smaller	Larger	K	Exact level (%)	K	Exact level (%)	K	Exact level (%)
17	18	103	90·4	94	95·1	76	99·1
17	19	110	90·0	100	95·1	82	99·0
17	20	116	90·3	106	95·2	87	99·0
17	21	122	90·5	112	95·2	92	99·1
17	22	129	90·2	118	95·2	97	99·1
17	23	135	90·5	124	95·2	103	99·0
17	24	142	90·2	130	95·2	108	99·0
17	25	148	90·4	136	95·2	113	99·1
18	18	110	90·3	100	95·3	82	99·0
18	19	117	90·2	107	95·1	88	99·0
18	20	124	90·1	113	95·2	93	99·1
18	21	131	90·0	120	95·1	99	99·0
18	22	137	90·5	126	95·2	105	99·0
18	23	144	90·4	133	95·0	110	99·1
18	24	151	90·4	139	95·2	116	99·0
18	25	158	90·3	146	95·0	122	99·0
19	19	124	90·4	114	95·0	94	99·0
19	20	131	90·5	120	95·3	100	99·0
19	21	139	90·1	127	95·3	106	99·0
19	22	146	90·3	134	95·2	112	99·0
19	23	153	90·4	141	95·2	118	99·0
19	24	161	90·1	148	95·2	124	99·0
19	25	168	90·2	155	95·1	130	99·1
20	20	139	90·4	128	95·1	106	99·1
20	21	147	90·2	135	95·2	113	99·0
20	22	155	90·1	142	95·3	119	99·0
20	23	162	90·4	150	95·1	126	99·0
20	24	170	90·3	157	95·2	132	99·0
20	25	178	90·2	164	95·3	139	99·0
21	21	155	90·3	143	95·1	119	99·1
21	22	163	90·4	151	95·0	126	99·1
21	23	171	90·4	158	95·2	133	99·1
21	24	180	90·1	166	95·2	140	99·0
21	25	188	90·2	174	95·1	147	99·0
22	22	172	90·2	159	95·1	134	99·0
22	23	180	90·5	167	95·1	141	99·0
22	24	189	90·3	175	95·2	148	99·1
22	25	198	90·1	183	95·2	156	99·0
23	23	190	90·0	176	95·0	149	99·0
23	24	199	90·1	184	95·2	156	99·1
23	25	208	90·1	193	95·1	164	99·0
24	24	208	90·3	193	95·2	165	99·0
24	25	218	90·1	202	95·2	173	99·0
25	25	228	90·1	212	95·1	181	99·0

For samples sizes where n_1 and n_2 are greater than the range shown in the table, a satisfactory approximation to the value of K can be calculated as

$$K = \frac{n_1 n_2}{2} - \left(N_{1-\alpha/2} \times \sqrt{\frac{n_1 n_2 (n_1 + n_2 + 1)}{12}}\right),$$

rounded up to the next higher integer value, where $N_{1-\alpha/2}$ is the appropriate value from the standard Normal distribution for the $100(1 - \alpha/2)$ percentile.

For $n_1 = 25$, $n_2 = 10$, and $\alpha = 0{\cdot}05$, for example, this calculation gives 71·3, which results in $K = 72$ for finding the 95% confidence interval, the same value as shown in the table.

For an explanation of the use of this table see chapter 8.

TABLE 6—Values of $K^\star$ for use in calculating confidence intervals for population medians in single samples or for differences between population medians for the case of two paired samples with sample size n from 6 to 50 and the associated exact levels of confidence, based on the Wilcoxon matched pairs signed rank sum distribution

Sample size (n)	Level of confidence					
	90% (approx)		95% (approx)		99% (approx)	
	$K^\star$	Exact level (%)	$K^\star$	Exact level (%)	$K^\star$	Exact level (%)
6	3	90·6	1	96·9	—	—
7	4	92·2	3	95·3	—	—
8	6	92·2	4	96·1	1	99·2
9	9	90·2	6	96·1	2	99·2
10	11	91·6	9	95·1	4	99·0
11	14	91·7	11	95·8	6	99·0
12	18	90·8	14	95·8	8	99·1
13	22	90·6	18	95·2	10	99·2
14	26	90·9	22	95·1	13	99·1
15	31	90·5	26	95·2	16	99·2
16	36	90·7	30	95·6	20	99·1
17	42	90·2	35	95·5	24	99·1
18	48	90·1	41	95·2	28	99·1
19	54	90·4	47	95·1	33	99·1
20	61	90·3	53	95·2	38	99·1
21	68	90·4	59	95·4	43	99·1
22	76	90·2	66	95·4	49	99·1
23	84	90·2	74	95·2	55	99·1
24	92	90·5	82	95·1	62	99·0
25	101	90·4	90	95·2	69	99·0
26	111	90·1	99	95·1	76	99·1
27	120	90·5	108	95·1	84	99·0
28	131	90·1	117	95·2	92	99·0
29	141	90·4	127	95·2	101	99·0
30	152	90·4	138	95·0	110	99·0
31	164	90·2	148	95·2	119	99·0
32	176	90·2	160	95·0	129	99·0
33	188	90·3	171	95·2	139	99·0
34	201	90·2	183	95·2	149	99·0
35	214	90·3	196	95·1	160	99·0
36	228	90·2	209	95·0	172	99·0
37	242	90·2	222	95·1	183	99·0
38	257	90·1	236	95·1	195	99·0
39	272	90·1	250	95·1	208	99·0
40	287	90·3	265	95·0	221	99·0
41	303	90·2	280	95·0	234	99·0
42	320	90·1	295	95·1	248	99·0
43	337	90·0	311	95·1	262	99·0
44	354	90·1	328	95·0	277	99·0
45	372	90·0	344	95·1	292	99·0
46	390	90·1	362	95·0	308	99·0
47	408	90·2	379	95·1	323	99·0
48	427	90·2	397	95·1	340	99·0
49	447	90·1	416	95·1	356	99·0
50	467	90·1	435	95·1	374	99·0

For sample sizes of n over 50, a satisfactory approximation to the value of $K^\star$ can be calculated as

$$K^\star = \frac{n(n+1)}{4} - \left(N_{1-\alpha/2} \times \sqrt{\frac{n(n+1)(2n+1)}{24}}\right),$$

rounded up to the next higher integer value, where $N_{1-\alpha/2}$ is the appropriate value from the standard Normal distribution for the $100(1-\alpha/2)$ percentile.

For $n = 50$ and $\alpha = 0{\cdot}05$, for example, this calculation gives 434·5, which results in $K^\star = 435$ for finding the 95% confidence interval, the same value as shown in the table.

For an explanation of the use of this table see chapter 8.

Notation

DOUGLAS G ALTMAN

Chapters 3 to 8 contain formulae for calculating confidence intervals. Repeated use is made of the mathematical notation explained below.

$\bar{x}$	the mean of a sample of observations, where the individual observations are denoted by x or x_i; it is pronounced "x bar." In some chapters we use y and d to denote sets of observations and $\bar{y}$ and $\bar{d}$ to denote their means.
p	the proportion with a certain characteristic in a sample of subjects.
SD (or s)	the standard deviation of a set of observations. It is a measure of their variability around the sample mean.
SE	the standard error of the sample mean or some other estimated statistic. It is a measure of the uncertainty of such an estimate and is used to derive a confidence interval in most of the chapters in this book. (The distinct uses and interpretation of the SD and SE are discussed in appendix 1, chapter 2. Note that the notation SE(b) or SE_b means "the standard error of b.")
Σ	the Greek capital letter sigma, denoting "sum of." Thus Σx means the sum of all the values of x. A more correct notation is $\sum_{i=1}^{n} x_i$, which means the sum of the n values of x_i, that is, $x_1 + x_2 + x_3 + \ldots + x_n$. The simpler notation

$\sum x$ is used when it is clear which items are being added together.

$\prod$ the Greek capital letter pi, denoting "product of." Thus $\prod x$ means the product of all the values of x. As with $\sum$ above, a fuller notation is $\prod_{i=1}^{n} x_i$, which is equal to $x_1 \times x_2 \times x_3 \times \ldots \times x_n$, but the shorter notation is used when the meaning is clear (see chapter 7).

(. . .) Brackets are used in formulae to clarify the structure and to indicate the correct method of calculation. The quantity inside brackets must always be calculated first. If there are brackets within brackets the inner quantity is evaluated first.

$\log_e x$ the logarithmic function giving the value y such that $x = e^y$, where e is the constant 2·718 281 . . . . $\log_e x$ is sometimes known as the natural logarithm of x, and an alternative notation is $\ln x$.

A key feature of the logarithmic transformation is that it is often successful in converting a non-Normal skewed distribution into an approximately Normal distribution (see chapter 3). Calculations, such as those to derive a confidence interval, can be performed using the log data and the results back transformed using the function e^x (see next entry).

e^x the exponential function denoting the inverse procedure to taking natural logarithms. It is sometimes called an antilogarithmic transformation. An alternative notation is $\exp(x)$.

n_i or N_i the sample size in the ith group of subjects.

$N_{1-\alpha/2}$; α; $100(1-\alpha)$ $N_{1-\alpha/2}$ represents a value from the "standard Normal distribution," which is the theoretical Normal distribution with mean 0 and standard deviation 1 (see fig 2.4). The subscript $1-\alpha/2$ represents the proportion of the distribution below the value $N_{1-\alpha/2}$. Thus $N_{0\cdot975}$ is the value from the standard Normal distribution below which lies the bottom 0·975 (or 97·5%) of the distribution.

The central $1-\alpha$, or $100(1-\alpha)\%$, of the distribution lies between $N_{\alpha/2}$ and $N_{1-\alpha/2}$. Because of the symmetry of the Normal distribution $N_{\alpha/2}=-N_{1-\alpha/2}$, so that the central $100(1-\alpha)\%$ of the distribution lies between $-N_{1-\alpha/2}$ and $N_{1-\alpha/2}$. For example, the central 0·95 (or 95%) of the Normal distribution lies between $-N_{0\cdot975}$ and $N_{0\cdot975}$—that is, between $-1\cdot96$ and $+1\cdot96$.

See also appendix 2, chapter 2.

$t_{1-\alpha/2}$ For some estimates, such as means and regression coefficients, the distribution of values from repeated sampling has a t distribution rather than a Normal distribution. For large samples the t distribution becomes nearly the same as the Normal distribution, but for small samples it has longer tails. As the tails of the distribution are relevant when calculating a confidence interval it is important to use the t distribution when appropriate. The logic behind the notation $t_{1-\alpha/2}$, however, is exactly as for the Normal distribution described in the preceding entry.

The t distribution, and hence the value of $t_{1-\alpha/2}$, is different according to the size of the sample(s) of data and is characterised by the "degrees of freedom." The method for calculating the relevant degrees of freedom is given in those chapters which make use of the t distribution.

In many cases both confidence intervals and hypothesis tests are calculated on the same data. It is important to remember that the value of the theoretical t distribution should be used for calculating a confidence interval, and not the observed value of the t statistic calculated in the hypothesis test.

P the probability value (or significance level) obtained from a hypothesis test. P is the probability of the data (or some more extreme data) arising by chance—that is, due to sampling variation only—when the null hypothesis is true. Hypothesis testing is discussed in chapters 2 and 9, but methods are not covered in detail in this book.

Index